Use of Chlorine Dioxide and Ozone for Control of Disinfection By-Products

The mission of the Awwa Research Foundation (AwwaRF) is to advance the science of water to improve the quality of life. Funded primarily through annual subscription payments from over 1,000 utilities, consulting firms, and manufacturers in North America and abroad, AwwaRF sponsors research on all aspects of drinking water, including supply and resources, treatment, monitoring and analysis, distribution, management, and health effects.

From its headquarters in Denver, Colorado, the AwwaRF staff directs and supports the efforts of over 700 volunteers, who are the heart of the research program. These volunteers, serving on various boards and committees, use their expertise to select and monitor research studies to benefit the entire drinking water community.

Research findings are disseminated through a number of technology transfer activities, including research reports, conferences, videotape summaries, and periodicals.

Use of Chlorine Dioxide and Ozone for Control of Disinfection By-Products

Prepared by:
Peter Zhou
Contra Costa Water District, Concord, California
and

Jeff Neemann
Black & Veatch Corporation, Kansas City, Missouri

Jointly sponsored by:
Awwa Research Foundation
6666 West Quincy Avenue, Denver, CO 80235-3098
and

California Energy Commission, PIER Program
Sacramento, CA
and

Contra Costa Water District
Concord, CA

Published by:

Awwa Research Foundation
Advancing the Science of Water®

IWA Publishing

DISCLAIMER

This study was funded by the Awwa Research Foundation (AwwaRF). AwwaRF assumes no responsibility for the content of the research study reported in this publication or for the opinions or statements of fact expressed in the report. The mention of trade names for commercial products does not represent or imply the approval or endorsement of AwwaRF. This report is presented solely for informational purposes.

Published by IWA Publishing, Alliance House, 12 Caxton Street, London SW1H 0QS, UK

Telephone: +44 (0) 20 7654 5500; Fax: +44 (0) 20 7654 5555; Email: publications@iwap.co.uk; Web: **www.iwapublishing.com**

AwwaRF report number 90981F. Originally published by AwwaRF for its subscribers in 2004.
IWA Publishing version published 2005.

ISBN 1 84339 879 6

CONTENTS

TABLES

FIGURES

FOREWORD

The Awwa Research Foundation is a nonprofit corporation that is dedicated to the implementation of a research effort to help utilities respond to regulatory requirements and traditional high-priority concerns of the industry. The research agenda is developed through a process of consultation with subscribers and drinking water professionals. Under the umbrella of a Strategic Research Plan, the Research advisory Council prioritizes the suggested projects based upon current and future needs, applicability, and past work; the recommendations are forwarded to the Board of Trustees for final selection. The foundation also sponsors research projects through the unsolicited proposal process; the Collaborative Research, Research Applications, and Tailored Collaboration programs; and various joint research efforts with organizations such as the U.S. Environmental Protection Agency, the U.S. Bureau of Reclamation, and the Association of California Water Agencies.

This publication is a result of one of these sponsored studies, and it is hoped that its findings will be applied in communities throughout the world. The following report serves not only as a means of communicating the results of the water industry's centralized research program but also as a tool to enlist the further support of the nonmember utilities and individuals.

Projects are managed closely from their inception to the final report by the foundation's staff and large cadre of volunteers who willingly contribute their time and expertise. The foundation serves a planning and management function and awards contracts to other institutions such as water utilities, universities, and engineering firms. The funding for this research effort comes primarily from the Subscription Program, through which water utilities subscribe to the research program and make an annual payment proportionate to the volume of water they deliver and consultants and manufacturers subscribe based on their annual billings. The program offers a cost-effective and fair method for funding research in the public interest.

A broad spectrum of water supply issues is addressed by the foundation's research agenda: resources, treatment and operations, distribution and storage, water quality and analysis, toxicology, economics, and management. The ultimate purpose of the coordinated effort is to assist water suppliers to provide the highest possible quality of water economically and reliably. The true benefits are realized when the results are implemented at the utility level. The foundation's trustees are pleased to offer this publication as a contribution toward that end.

Edmund G. Archuleta, P.E.
Chair, Board of Trustees
Awwa Research Foundation

James F. Manwaring, P.E.
Executive Director
Awwa Research Foundation

ACKNOWLEDGMENTS

The authors of this report are indebted to the following water utilities and individuals for their cooperation and participation in this project:

Contra Costa Water District, Concord, California
Diablo Water District, Oakley, California

The advice to the project team and advocacy for the project provided by the Awwa Research Foundation Project Manager, Dr. Kenan Ozekin, and thoughtful guidance of the Project Advisory Committee are gratefully acknowledged. The Project Advisory committee consisted of Kevin Dixon, New Jersey American Water Company, Dr. Issam Najm, Water Quality & Treatment Solutions, Robert Riley, El Paso Water Utilities.

The contributions and guidance of the California Energy Commission Project Manager Shahid Chaudhry are also acknowledged. The authors wish to thank John Murphy, Global Energy Partners, for his contributions in preparing Chapter 9, energy uses and impacts. The authors also acknowledge the advice and guidance of Dale Newkirk of Damon S. Williams Associated, LLC, and Keith Carns and Dave Perkins of Global Energy Partners.

The authors wish to thank ERCO Worldwide (formerly Sterling Pulp Chemicals Ltd.), Water Technologies Division and International Dioxcide, Inc. for their generous in-kind contributions throughout the project. The authors also wish to acknowledge Peter Deacon, John Christie, Dr. Georges Ranger, Dave Dean, Ron Schaible, Dr. Yuri Omelchenko, Ty Tran, and Tim Mutto of ERCO Worldwide and Peter Mischuk of International Dioxcide Inc. for their technical and on-site assistance during the study.

The authors wish to acknowledge and thank the Randall-Bold Operations Staff for their support and hard work during the study. They are: Steve Bow, Lee Burnett, Luis Ching, Ron Gardner, John Harvey, Daniel Hewes-Hartman, Lyall Miller, Matthew Novak, Jerry Novero, John Parsons, and Werner Schmid. Roger Swartfager provided numerous technical contributions and was instrumental in operating the plant during the study. The authors also wish to thank Dave Huey, Pat Panus, and Steve Hickman for their input and continuous support of the study by always allowing one more test.

The quality and quantity of the data from this study could not have been produced without the assistance of the CCWD Laboratory Staff. The authors wish to thank Larry McCollum, Jean Zacher, and Joe Gustino for their technical contributions and support during the study. The authors also wish to thank Gail Ross and Wendell Bashaw for their hard work.

The authors wish to acknowledge the technical contributions and assistance of Dr. Robert C. Hoehn. Dr. Hoehn provided valuable advice during the project and was a major contributor in preparing Chapter 2. The authors wish to thank Nick Burns, Bob Hulsey, Bruce Long, Perri Standish-Lee, and Joe Wong of Black & Veatch for their assistance during the project. The authors would also like to acknowledge the technical contributions to regulatory history of this report and on-site assistance of Mr. Tai Tseng, McGuire Environmental Consultants. The authors also wish to thank Dr. Michelle Frey, McGuire Environmental Consultants, for her assistance and review, and Dr. Don Gates and Dr. Gil Gordon for their fruitful discussions during the project.

EXECUTIVE SUMMARY

Control of disinfection byproducts (DBPs) in drinking water is becoming increasingly critical as regulatory requirements are calling for higher and higher levels of disinfection while at the same time mandating lower maximum contaminant levels (MCLs) of DBPs. Many utilities have switched to alternative disinfectants such as ozone and chlorine dioxide (ClO_2) to minimize chlorinated DBPs such as trihalomethanes (THMs) and haloacetic acids (HAAs); however, ozone and ClO_2 form their own DBPs, some of which are regulated and some are not. There is a potential synergy in using a combination of ozone and ClO_2 to increase the disinfection effectiveness while reducing DBPs, but few utilities have studied this synergy. On the other hand, with possible ozone demand reduction due to addition of ClO_2 before ozone, it is expected potential energy saving for such ozone dose reduction. With these potential benefits in mind, the Contra Costa Water District (CCWD) undertook a full-scale demonstration study using ClO_2 and ozone to evaluate the potential benefits of combining these two powerful oxidants.

The full-scale demonstration was conducted at the 40 mgd Randall-Bold Water Treatment Plant (WTP). The Randall-Bold WTP is a direct filtration plant that uses preozonation (before coagulation) to enhance the filtration process and postozonation (after filtration) for primary disinfection. The filters are deep bed granular activated carbon (GAC) and sand filters, with 80 inches of GAC and 40 inches of sand. Chloramines are used for secondary disinfection by adding the ammonia before the postozonation contactor and the chlorine after the postozonation contactor. During the study, ClO_2 was added to the raw water ahead of the preozonation contactor.

The testing component of the study was divided into three phases that served as building blocks to take the evaluation from bench-scale ClO_2 demand studies to long-term full-scale operation using ClO_2 in combination with ozone. The first phase consisted of screening different dosages of the combination of ClO_2 and ozone for further testing. During the second phase, selected dosage combinations of ClO_2 and ozone were evaluated by short-term tests of two days' duration, which is longer than the typical filter run of 24 hours at the Randall-Bold WTP. The third, and final, phase consisted of three-week long-term tests to evaluate the project objectives as well as the operational challenges that might be encountered when using the combination of ClO_2 and ozone. Three preoxidation conditions were tested: 1.5 mg/L of ClO_2 alone, the combination of 1.0 mg/L of ClO_2 and 0.5 mg/L of ozone, and 1.0 mg/L of ozone alone. Each condition was evaluated for three weeks, with one replicate, for a total of 18 weeks of full-scale operation. During the third phase, energy consumption data were collected to evaluate the potential energy saving under different scenarios. The data are also used to further estimate California statewide energy saving under these scenarios and other conditions. In addition, special full-scale testing was conducted to evaluate bromate mitigation techniques such as preoxidation with ClO_2, pH adjustment, and ammonia addition.

OBJECTIVES

The principal research objective, stated broadly, was to evaluate the benefits of using ClO_2 in combination with ozone. The specific objectives for the evaluation were to determine the extent to which ClO_2 preoxidation of the raw water (1) reduced the ozone demand/dose in the preozonation and postozonation contactors, (2) reduced the bromate formation by reducing the ozone dose, (3) reduced the energy consumption, (4) resulted in chlorate ion formation by ozone's oxidation of chlorite ion, (5) maintained the filtration benefits of preozonation, and (6) reduced THM and HAA formation potentials in finished water. Additional objectives of the testing included a full-scale evaluation of three bromate mitigation techniques at the Randall-Bold WTP. The techniques investigated included (1) raw-water preoxidation with ClO_2 as a means of reducing the overall ozone demand during preozonation (where most of the bromate is formed), (2) addition of sulfuric acid to raw water for pH reduction before preozonation, and (3) addition of ammonia to raw water ahead of the preozonation contactor as a means of forcing a higher percentage of bromamine formation during the reaction of ozone with bromide ion. The comparison of the three mitigation techniques was limited to determination of the applicability of each method to the Randall-Bold WTP.

RESULTS

Bromate Formation

During the long-term full-scale testing, bromate formation was monitored to evaluate the possible benefits of ClO_2 preoxidation. The results showed that ClO_2 preoxidation reduced the finished water bromate concentration when compared to the control tests done without ClO_2 (Table ES.1). Tests with ClO_2 preoxidation resulted in bromate concentrations less than the detection limit of 1 ppb in all the samples after preozonation, and in average finished water bromate concentrations less than 2 ppb. On average, the two test conditions with ozone alone, however, resulted in bromate concentrations of 6.2 and 3.9 ppb in the finished water.

Table ES.1
Impact of preoxidation on postozonation dose and bromate formation

Phase	Average raw water bromide (ppb)	ClO_2 dose (mg/L)	Pre-O_3 dose (mg/L)	Average post-O_3 dose (mg/L)	Average post-O_3 CT (mg·min/L)	Average* bromate finished water (ppb)
3A	265	1.5	-	1.22	4.26	1.5
3B	257	1.0	0.5	0.78	3.99	1.9
3C	305	-	1.0	0.54	3.94	6.2
3D	343	1.5	-	1.53	4.97	0.6
3E	316	1.0	0.5	1.16	5.16	2.0
3F	255	-	1.0	1.13	5.14	3.9

* Concentrations below the detection limit were treated as zero in the average calculation

Results from the special testing showed that preoxidation with ClO_2 and pH adjustment were very effective in reducing the bromate formation at an ozone dose of 2 mg/L. Ammonia addition was effective, but to a lesser extent. The July tests showed that using ozone alone would result in bromate concentrations exceeding the MCL (10 ppb), but either preoxidation with 1.0 mg/L ClO_2 or lowering the pH value to 6.0 reduced bromate concentrations to half the MCL (Table ES.2). The results also showed that substituting sodium chlorite ($NaClO_2$) for ClO_2

addition also reduced bromate formation, which strongly indicates that chlorite plays an important role in reducing bromate formation when ClO_2 is added. However, when the ozone dose was increased to 4.0 mg/L, the effectiveness of preoxidation with ClO_2 or chlorite was decreased.

Table ES.2
Average bromate formation during July 2002 bromate mitigation testing

Condition	Ozone Dose (mg/L)	Bromide (ppb)	TOC (mg/L)	UV254 (1/cm)	Average Bromate (ppb)
Ozone only	2	160	2.0	0.085	24.7
0.5 mg/L ClO_2	2	180	2.1	0.095	14.3
1.0 mg/L ClO_2	2	180	2.1	0.095	**5.2**
0.7 mg/L Chlorite	2	180	2.0	0.085	**5.3***
pH 6.5	2	180	2.0	0.085	13.3
pH 6.0	2	180	2.0	0.085	**5.6**
0.25 mg/L NH3-N	2	210	2.0	0.087	19.7
0.5 mg/L NH3-N	2	210	2.0	0.087	16.9
Ozone only	4	180	2.1	0.095	60.3
0.5 mg/L ClO_2	4	180	2.1	0.095	62.3
1.0 mg/L ClO_2	4	160	2.0	0.085	57.7
0.7 mg/L Chlorite	4	180	2.0	0.085	51.3

*Represents data from third sample, not the average of three samples

Chlorite and Chlorate Formation

The chlorite and chlorate concentrations were monitored at several points in the plant to determine the amount of chlorite that formed after ClO_2 addition and the amount that was oxidized to chlorate by ozone. The results showed that there were three chlorite removal processes in the treatment train and that effluent chlorite concentrations were typically below the detection limit of 10 ppb. Both the preozonation and postozonation converted a majority of the chlorite to chlorate. The GAC filters typically removed about half of the chlorite present in the filter influent; however, unlike ozone, the GAC removed the chlorite without causing the chlorate concentration to increase.

Impacts on Coagulation and Filtration

An important factor in substituting preoxidation with ClO_2 for ozone is that it does not have an adverse impact on the performance of the coagulation and filtration process. To evaluate the potential impact of substituting ClO_2, the coagulant doses, filtered water turbidities, TOC removal, and DBP formation potential were monitored during the long-term testing. The results showed that although the preoxidation scheme was rotated among three conditions, the coagulant and polymer dosages did not need to be substantially changed and there was little difference in filter effluent turbidities. In terms of organics removal, the removal of TOC and DBP formation potential for the three preoxidation conditions were also nearly the same. Therefore, it appears that ClO_2 could be substituted for ozone without adverse impact on the coagulation and filtration process.

Operational Impacts

An important consideration related to the use of ClO_2 is issues that affect daily operations; several were identified during the long-term testing program. One issue was that the ClO_2 required additional daily attention for the operators to tune the generator, perform process residual measurements, and the daily plant effluent compliance sampling and analysis. Another issue was using ClO_2 required additional monthly distribution sampling for chlorite. Also, because of the interaction of ClO_2 and chlorite with ozone, the operators had to monitor postozonation residuals more closely. The California action level (AL) for chlorate will limit the ClO_2 dose operations uses, because this study showed a majority of the chlorite formed after ClO_2 addition was converted to chlorate.

Energy Analysis

Using ClO_2 in combination with ozone will produce a modest reduction in the energy demand and electricity consumption at water treatment plants. Given the benefits of water quality improvement, moderate energy saving and increased complexity associated with using multiple chemicals, planners should weigh water quality impacts more heavily than energy savings. The optimization of ozone generation and use can substantially reduce energy use. From the standpoint to lowering statewide energy demand, the projected savings of replacing preozonation with chlorine dioxide (based on the results of this study and extrapolated to the entire state) are moderate. If energy demand reduction is the principal goal, the California Energy Commission should not commit extensive resources to implementing this change across the state. On the other hand, with potential water quality benefits of combining the two disinfectants/oxidants, the benefit of moderate energy saving is certainly an added bonus.

CONCLUSIONS

- Although preoxidation with ClO_2 forms chlorite, and the chlorite creates ozone demand for the postozonation contactor, the total ozone dose (preozonation plus postozonation) was reduced.
- The data from this study showed that preoxidation with ClO_2 substantially reduced the bromate formation. In addition to reducing the bromate formation by reducing the ozone dose through substitution with ClO_2, an unexpected result was that ClO_2 preoxidation also reduced the bromate formation at a constant ozone dose. Special testing showed that chlorite, which is formed after ClO_2 addition, plays an important role in the bromate formation reduction.
- Preoxidation with ClO_2 and pH adjustment to a value of 6.0 were effective methods to reduce the bromate formation in the preozonation contactor. Ammonia addition before

the preozonation contactor was effective, but to a lesser extent than ClO_2 and pH adjustment.

- The data from this study showed that ozone oxidized the chlorite to chlorate, so that the plant effluent chlorite concentration was below the detection limit. The testing also showed that some of the chlorite was removed by the GAC, while chlorate neither increased nor decreased through the GAC filters.
- Preoxidation with ClO_2 alone or ClO_2 in combination with ozone required similar coagulant dosages and produced similar filter effluent turbidity when compared to using ozone alone.
- The data showed that there was not a substantial difference in DBP formation potential between the three preoxidation conditions tested during the long-term phase of this study.
- Implementing ClO_2 in the treatment process at Randall-Bold WTP added complexity to the operations staff that included additional sampling, time consuming amperometric analyses, and the need to pay close attention to the postozonation dose because of the ozone demand created by chlorite.
- Using ClO_2 in combination with ozone resulted in a modest reduction in the energy demand and electricity consumption at the Randall-Bold WTP. The projected statewide energy savings (based on the results of this study and extrapolated to all California water utilities using ozone) are moderate.
- Optimization of ozone generation and ozone dose can have a large impact on the energy uses at a water treatment plant.

CHAPTER 1
INTRODUCTION

Control of disinfection byproducts (DBPs) in drinking water is becoming increasingly critical as regulatory requirements are calling for higher and higher levels of disinfection while at the same time mandating lower maximum contaminant levels (MCLs) of DBPs. Many utilities have switched to alternative disinfectants such as ozone and chlorine dioxide (ClO_2) to minimize chlorinated DBPs such as trihalomethanes (THMs) and haloacetic acids (HAAs); however, ozone and ClO_2 form their own DBPs, some of which are regulated and some are not. There is a potential synergy in using a combination of ozone and ClO_2 to increase the disinfection effectiveness while reducing DBPs, but few utilities have studied this synergy. With these potential benefits in mind, the Contra Costa Water District (CCWD) undertook a full-scale demonstration study using ClO_2 and ozone to evaluate the potential benefits of combining these two powerful oxidants.

The full-scale demonstration was conducted at the 40 mgd Randall-Bold Water Treatment Plant (WTP). The Randall-Bold WTP is a direct filtration plant that uses preozonation (before coagulation) to enhance the filtration process and postozonation (after filtration) for primary disinfection. The filters are deep bed granular activated carbon (GAC) and sand filters, with 80 inches of GAC and 40 inches of sand. Chloramines are used for secondary disinfection by adding the ammonia before the postozonation contactor and the chlorine after the postozonation contactor. During the study, ClO_2 was added to the raw water ahead of the preozonation contactor.

The testing component of the study was divided into three phases that served as building blocks to take the evaluation from bench-scale ClO_2 demand studies to long-term full-scale operation using ClO_2 in combination with ozone. The first phase consisted of screening different dosages of the combination of ClO_2 and ozone for further testing. During the second phase, selected dosage combinations of ClO_2 and ozone were evaluated by short-term tests of two days' duration, which is longer than the typical filter run of 24 hours at the Randall-Bold WTP. The third, and final, phase consisted of three-week long-term tests to evaluate the project objectives as well as the operational challenges that might be encountered when using the combination of ClO_2 and ozone. Three preoxidation conditions were tested: 1.5 mg/L of ClO_2 alone, the combination of 1.0 mg/L of ClO_2 and 0.5 mg/L of ozone, and 1.0 mg/L of ozone alone. Each condition was evaluated for three weeks, with one replicate, for a total of 18 weeks of full-scale operation. In addition, special full-scale testing was conducted to evaluate bromate mitigation techniques such as preoxidation with ClO_2, pH adjustment, and ammonia addition.

CHLORINE DIOXIDE AND OZONE

Individually, ClO_2 and ozone have been used extensively for water treatment in both the United States and around the world. Uses of ClO_2 include disinfection, oxidation of iron and manganese, taste and odor control, enhancement of coagulation and filtration, reducing chlorinated DBPs, and control of biological growth in open treatment basins. Hoehn et al. (2000) summarizing the Information Collection Rule (ICR) data reported that 34 of 383 surface water plants used ClO_2 for treatment; however, they noted that many small plants that use ClO_2 were not included in the ICR. Like ClO_2, ozone has many uses that include disinfection, oxidation of iron and manganese, taste and odor control, enhancement of coagulation and filtration, reducing chlorinated DBPs, and oxidation of hydrogen sulfide. Rice et al. (1998) at the time, estimated there were 264 water plants using ozone in the United States and that number has likely grown over the last several years. Although both ClO_2 and ozone have numerous applications, few utilities have used a combination of the two and evaluated the advantages and disadvantages of doing so.

Adding ClO_2 and ozone at the same time would not be effective because ozone reacts rapidly with ClO_2 to form chlorate. In essence, the benefits of each would be cancelled out if they were added to the same water at the same location. Therefore, this study evaluated sequential application of ClO_2 and ozone; or specifically, the addition of ClO_2, followed by a reaction period, and then the addition of ozone. A desired outcome of ClO_2 addition to the raw water was to lower or even eliminate the preozonation dose. It is hoped that this method of combining oxidants will allow the use of lower oxidant dosages than by using each oxidant alone, which would produce savings in chemical costs and energy use, as well as reducing DBP formation.

DISINFECTION BYPRODUCT FORMATION

A balance must always be struck between disinfection and the formation of disinfection byproducts. Chlorinated DBPs such as THMs and HAAs have long been a concern for the water treatment industry because some are carcinogens, which has helped fuel the switch over to alternative disinfectants such as chloramines, ClO_2, ozone, and ultraviolet (UV) light. However, these alternative disinfectants have their own DBPs. While chloramines substantially reduce the formation of THMs and HAAs, recent work suggests that monochloramine can cause the formation of N-Nitrosodimethylamine (NDMA) (Mitch and Sedlak 2002). The use of ClO_2 will result in the formation of chlorite, which has an MCL of 1.0 mg/L, and chlorate, which in California has an action level of 0.8 mg/L. Ozonation of bromide containing waters can cause the formation of bromate, which has been identified as a probable carcinogen and has an MCL of 0.010 mg/L. UV is an evolving technology in water treatment and research to date has shown no direct DBP formation at the doses used for disinfection, except for some formation of nitrite from nitrate (von Sonntag et al. 1993).

Bromate

Bromate formation is a concern for utilities that use ozone, but it can be controlled by optimizing ozone use, by reducing ozone demand with preoxidation, by pH adjustment, and by adding ammonia. Since bromate formation is a function of the ozone dose, optimizing the ozone feed system through better controls or refining the ozone disinfection credit calculation can lower the dose and hence the bromate formation. Preoxidation with chlorine or ClO_2 can work in a similar manner by reducing the ozone demand and consequently the ozone dose. Adjustment of pH can drive the bromide chemistry to form more hypobromous acid, instead of hypobromite, which reduces the bromate formation. Ammonia addition can reduce bromate formation by forming bromamines, which are similar to chloramines, and when the bromide is "tied up" with the ammonia, it does not react with ozone to form bromate. This study focused on using ClO_2 preoxidation to reduce the ozone dose and the bromate formation. In addition, special testing was conducted to compare preoxidation with ClO_2 to the two most common techniques for bromate control, pH adjustment and ammonia addition.

Chlorite/Chlorate

When ClO_2 is added in water treatment, it results in the formation of both chloride and chlorite. Typical ranges for the chlorite concentration after ClO_2 addition are 30 to 70 percent of the reacted ClO_2 dose (Gates 1998). Chlorate can result from inefficient ClO_2 generation or from the reaction of chlorite with free chlorine. In this study, chlorate formation was more prevalent because ozone reacts rapidly with chlorite to form chlorate. The chlorite and chlorate concentrations are important because chlorite has an MCL of 1.0 mg/L and the California Department of Health Services recently established an action level for chlorate at 0.8 mg/L.

The chlorite can be removed with reduced-iron (ferrous) and reduced-sulfur compounds. Granular activated carbon (GAC) can also remove chlorite by reduction to chloride, though its capacity is limited (Hoehn et al. [N.d.]). Chlorate is not removed by typical water treatment processes, so advanced treatment with reverse osmosis or ion exchange would be required for its removal. The most common way to control the chlorite and chlorate concentrations is to limit the ClO_2 dose to around 1.5 mg/L so the resulting chlorite and chlorate concentrations do not exceed the allowable limits. This study used a combination of limiting the ClO_2 dose and relying on the GAC filters to remove chlorite.

ELECTRIC POWER IMPLICATIONS OF THE PROCESS CHANGE

In addition to the differences in their impact on bromate formation, ClO_2 and ozone have very different energy use characteristics. Chlorine dioxide requires significantly less electrical energy, even when taking into account the energy used to generate the chlorine and sodium chlorite. In fact, with liquid oxygen as the feed ozone requires a total of 10 to 20 kWh of electricity per pound of ozone generated while ClO_2 synthesis requires about 6.5 kWh per pound generated. Further, given the reactivity of ozone (and its subsequent reactions with naturally-occurring organic matter), applied ozone doses are proportionally greater than what is typically needed for ClO_2. In practice, energy consumption associated with ozone use is higher than the consumption associated with ClO_2. So, a separate goal of this study was to assess the energy impact of switching from ozone to ClO_2 at the Randal-Bold WTP.

Given the electric use characteristics of ozone and ClO_2, the implications of instituting this process change statewide were also considered. The project team estimated before the study that if the process change assessed in this study were to be implemented at all 33 water treatment plants in California currently using ozone, potential power savings would approach 4 MW. Based on this estimate, the team petitioned the California Energy Commission's Public Interest Energy Research (PIER) program for additional funding to evaluate the possible statewide impact.

The CEC PIER program is dedicated to public interest energy research development and demonstration. The program's mission is to conduct research that improves the quality of life for Californians by providing environmentally sound, safe, reliable and affordable energy services and products. Among the programs' five specific focus areas is end-use energy efficiency, so CEC staff were attracted to how this study was to assess one way California water treatment plants might optimize both treatment and energy consumption. After the panic precipitated by California's energy crisis in 2001, the CEC staff were particularly keen on methods to reduce statewide electrical demand. If the estimated power savings were achieved, implementing this process statewide would meet several goals of the PIER program, including increasing the value of electricity used in California, reducing the environmental impacts of its use, and enhancing California's economy. Given the significant potential power reduction, and the urgency precipitated by the power crisis of 2001, the Energy Commission funded an assessment of the statewide implications to California's electricity demand as an outgrowth of this study.

This energy assessment was conducted on two levels: a local approach and a statewide survey. The first assessment considered the impact of the process change on electricity use and demand at the Randall-Bold WTP. Another component of this assessment was to estimate the impact on the state's electricity demand by assuming the implementation of a similar process change at all water treatment plants in California.

RESEARCH GOALS AND OBJECTIVES

The principal research objective, stated broadly, was to evaluate the benefits of using ClO_2 in combination with ozone. The specific objectives for the evaluation were to determine the extent to which ClO_2 preoxidation of the raw water

(1) reduced the ozone demand/dose in the preozonation and postozonation contactors,
(2) reduced the bromate formation by reducing the ozone dose,
(3) reduced the energy consumption both at Randall-Bold WTP and statewide,
(4) resulted in chlorate ion formation by ozone oxidation of chlorite ion,
(5) maintained the filtration benefits of preozonation, and
(6) reduced THM and HAA formation potentials in finished water.

Additional objectives of the testing included a full-scale evaluation of three bromate mitigation techniques at the Randall-Bold WTP. The techniques investigated included (1) raw-water preoxidation with ClO_2 as a means of reducing the overall ozone demand during preozonation (where most of the bromate is formed), (2) addition of sulfuric acid to raw water for pH reduction before preozonation, and (3) addition of ammonia to raw water ahead of the preozonation contactor as a means of forcing a higher percentage of bromamine formation during the reaction of ozone with bromide ion. The comparison of the three mitigation techniques was limited to determination of the applicability of each method to the Randall-Bold WTP.

REPORT ORGANIZATION

The following report presents a summary of the full-scale demonstration of using a combination of ClO_2 and ozone. Chapter 2 presents background information on ClO_2 and on the Randall-Bold WTP, where the testing was conducted. The materials and methods used during the testing, as well as the testing protocol that was followed, are discussed in Chapters 3 and 4. Results are summarized in Chapters 5 through 8 for Phases 1, 2, and 3 and for the bromate mitigation testing. The energy uses and the potential impacts of using ClO_2, both at the Randall-Bold WTP and statewide, are listed in Chapter 9. Finally, the conclusions from the testing, the recommendations based on the results, and the possible future work are presented in Chapter 10.

CHAPTER 2
BACKGROUND

CURRENT STATE OF KNOWLEDGE

The following section describes the current state of knowledge of ClO_2. This section describes ClO_2 use in the United States, its uses for drinking water, its benefits associated with this project, the regulatory history for drinking water, an overview of ClO_2 generation chemistry, and historic energy uses in drinking water.

Chlorine Dioxide Use in the United States

Two United States Environmental Protection (USEPA) documents contain information about ClO_2 use in the United States: the Community Water System Survey (USEPA 1997) and the Information Collection Rule (ICR) (*Federal Register 1996)*, and a data summary appeared in Appendix A of the *Alternative Disinfectant and Oxidants Guidance Manual* (*USEPA 1999).* The ICR included three categories of public water supplies, namely 1) those serving treated surface water to more than 100,000 people, 2) those serving water from groundwater sources only to more than 100,000 people, and 3) those serving at least 50,000 people but less than 100,000 and using ground water sources. Hoehn et al. (2000) presented a synopsis as follows:

- 81 percent of all community systems in 1995 provided some form of treatment to all or a portion of their water sources, and 6.3 percent of these reported ClO_2 use. (A system may contain more than one water treatment plant.)
- 73.7 percent of 529 systems (390 total) that were included in the ICR treat surface water while 25.5 percent (135 plants) treat groundwater. Only two systems (0.3%) treating a combination of surface and groundwater use ClO_2.
- 34 treatment plants included in the ICR out of a total of 383 that use surface water apply ClO_2. Twenty percent of these treat 11-50 mgd, another 11 percent treat 51-100 mgd, and 3 percent treat more than 100 mgd.
- ClO_2 is used by only one percent of 144 ICR systems that treat ground water, and all of these treat from 11-50 mgd.
- 31 of 527 ICR treatment plants (both ground water and surface water supplies) apply both Cl_2 and ClO_2, 18 apply both chloramine and ClO_2, and one applies both ozone and ClO_2. None of the 527 plants uses sodium hypochlorite and ClO_2 in combination. In addition, none of the groundwater systems applying ClO_2 use an additional oxidant.

Hoehn et al. (2000) also pointed out that many smaller plants (i.e. those serving less than 50,000 people) that use ClO_2 were not included in the ICR, and, therefore, the data listed above are misleading in that they are underestimates of the true number of ClO_2 users in the United States.

Major Uses of Chlorine Dioxide for Drinking Water Treatment

Even though ClO_2 offers a variety of benefits for drinking water treatment, many state regulators and consultants in the past have been reluctant to recommend its use. In some instances, the reluctance is borne of either bad experiences or unfavorable results, while in others it is the result either of misinformation about the nature of ClO_2 or unfamiliarity with its potential benefits. Hoehn et al. (2000) surveyed state regulators regarding ClO_2 use in their respective states and found that 15 states have no utilities that use it. Table 2.1 summarizes the reasons regulators in states where ClO_2 is used gave for using the oxidant.

Table 2.1
Reasons given for ClO_2 use

Reason for Using ClO_2	States Reporting Specific Use
Preoxidation (iron, manganese, sulfide, etc.)	AL, AR, GA, MO, NE, NM, SC, TN, TX* VA, WA, WY
THM and HAA Reduction	AZ, AR, CO, FL, GA, IL, IA, KS, KY, LA, ME, MA, MI, NH, NY, NC, OH, PA, RI, SD, TX, VA, WA
Disinfection	AR, CO, GA, IN, IA, LA, ME, MA, MI, NH, NY, NC, NC, OH, PA, RI, SD, TX, VA
Post Disinfection* (added prior to distribution)	IA, ME, NH, VA
Odor Control	CT, GA, LA, MI, MO, NM, OH, OK, PA, TX
Color Control	MI, MO, SC
Control of Biological Growth in Basins	CA (in open reservoir), CO, MO

Source: Hoehn *et al.*, 2000.

*At one time, Texas permitted application of ClO_2 prior to distribution, but odor complaints forced the Public Drinking Water Section of the Natural Resources Conservation Commission to withdraw permission for such applications.

Additional Benefits of Chlorine Dioxide Associated with this Project

While ClO_2 is a strong oxidant, it is not as strong as ozone and, therefore, will not oxidize bromide to bromate. This characteristic is another attractive feature of ClO_2, especially in areas where bromide concentrations are elevated. In many locations, high bromide concentrations in the raw water make ozone use impractical or impossible because of the constraints imposed by the bromate MCL (0.010 mg/L, *Federal Register,* 1998). According to Reckhow (1999), bromate formation is approximately linear when the ozone dose exceeds 0.5 mg/L, but it is inversely related to the dissolved organic carbon concentration at a fixed ozone dose. He also noted that lowering pH and minimizing aqueous ozone residuals are among the most effective bromate control strategies. Chlorite ion, like bromide, can be oxidized by ozone, forming chlorate ion (Hoigne, 1998).

One potential advantage of chlorite use not often discussed is that chloramine residuals and ammonia-nitrogen concentrations persist longer in systems that have applied ClO_2, presumably a result of chlorite inhibition of nitrifying bacteria (McGuire et al. 1999). The authors found that nitrifying bacteria activity could be suppressed by chlorite at concentrations as low as 0.05 mg/L. Protection is provided against nitrification and, thus, biofilm development can be controlled in distribution systems where autotrophic, nitrifying bacteria utilize ammonia derived from chloramines. This benefit would certainly apply to CCWD because they use chloramines for the distribution system disinfectant residual.

Regulatory History of Chlorine Dioxide Use in Water Treatment

Chlorine Dioxide as a Disinfectant

Interest in ClO_2 as a disinfectant and oxidizing agent increased in the 1970s and 1980s with the promulgation of the first Disinfection Byproducts (DBP) rule to control total trihalomethanes (TTHMs) in drinking water. In contrast with chlorine, the reactions of ClO_2 with naturally occurring organic matter and bromide do not result in the formation of appreciable amounts of TTHMs and produce lower levels of other halogenated organic DBPs. Its use was further stimulated when in 1983 the U.S. EPA listed ClO_2 as an alternative or supplemental oxidant-disinfectant that was one of the more suitable treatment technologies for control of TTHMs.

Although ClO_2 did not produce significant amounts of TTHMs, ClO_2 does produce chloride, chlorite, and chlorate during the disinfection process. In addition, chlorite can be further oxidized to chlorate during treatment or in the distribution system. Thus, in 1983, EPA established a guideline that the combined concentrations of ClO_2 and its two major byproducts, chlorite and chlorate, should not exceed 1 mg/L in treated drinking water.

In recent years, ClO_2 use has increased throughout many portions of the U.S. and Europe, with utilities limiting their dose to not exceed the federal guideline for the sum of the oxyhalogen species. However, because chlorite and chlorate have been associated with hemolytic anemia and methemoglobinemia, the California Department of Health Services (DHS) had restricted the use of the oxidant. Between 1986 and 1990, DHS exercised what it called a moratorium on the use of ClO_2 in situations where treated water would be used by facilities performing renal dialysis. According to DHS, "the basis of the moratorium was the lack of information on the effects of ClO_2 and the oxychlorine byproducts, chlorite and chlorate, on renal dialysis patients." Granular activated carbon, typically used in many renal dialysis facilities, does not control chlorate exposure.

Based on a toxicity review undertaken by the Office of Environmental Health and Hazard Assessment (OEHHA)'s predecessor, the Hazards Evaluation Section, and information provided to DHS by the USEPA, DHS went forward with a 1990 ClO_2 policy designed to allow its use in drinking water and at the same time protect individuals undergoing dialysis from exposure to levels of ClO_2 and oxychlorine byproducts that would cause oxidative damage to red blood cells. With this approach, DHS believed no additional treatment would be required at dialysis facilities.

The 1990 California State ClO_2 policy (Policy Memo 90-1, June 12, 1990) established Action Levels (ALs) for ClO_2 at 0.02 mg/L, chlorate at 0.2 mg/L, and chlorite at 0.02 mg/L. The ALs for chlorite and chlorate were established at what DHS considered to be the analytical detection limits at the time. The policy allowed the use of ClO_2 in drinking water treatment providing that no detectable levels of these chemicals exceeding the ALs would enter the distribution system and that finished water monitoring would be conducted.

The conservative ALs established in the 1990 State ClO_2 policy effectively precluded the use of ClO_2 for drinking water disinfection in California. Later, however, DHS revised their policy in November 1998 (Policy Memo 98-005) to reflect federal regulatory levels for ClO_2 and chlorite that were being promulgated in the U.S. EPA Stage 1 Disinfectants/Disinfection ByProducts (D/DBP) Rule. In December 1998, the USEPA promulgated a ClO_2 maximum residual disinfectant level (MRDL) of 0.8 mg/L and a chlorite maximum contaminant level (MCL) of 1.0 mg/L in the D/DBP Rule (*Federal Register* 1998). Both the chlorite MCL Goal (MCLG) and the ClO_2 MRDL Goal (MRDLG) that had originally been proposed at lower levels were increased to 0.8 mg/L because ClO_2 is reduced to chlorite and chloride when it reacts with reducing substances in water.

Monitoring requirements when ClO_2 is being used include the following: 1) daily analyses of ClO_2 and chlorite ion at the entrance to the distribution system by either amperometric titration or the DPD method and 2) monthly chlorite analyses at three points in the distribution system, one near the first customer and one each at points representing the midpoint and maximum residence times. Additional sampling is required if either of the regulatory limits is exceeded.

The 1998 California State ClO_2 policy (Policy Memo 98-005) supercedes the 1990 policy (Policy Memo 90-1) and recommends that water systems not exceed a ClO_2 concentration of 0.8 mg/L and a chlorite concentration of 1.0 mg/L. The policy includes a Notification of Known Sensitive Populations, as follows: "Since health effects studies have demonstrated that chlorate can cause oxidative damage to red blood cells and could cause methemoglobinemia and hemolytic anemia in sensitive individuals such as those undergoing dialysis and those with certain enzyme deficiencies, dialysis centers should be notified when a water system is proposing to change to chlorine dioxide."

Since compliance with the Stage 1 D/DBP rule for large surface-water systems became effective in January 2002, it is not clear that the 1998 CDHS policy on ClO_2 is now superceded itself by regulation, even though California is not expected to promulgate its own version of Stage 1 D/DBP Rule until end of 2002.

Chlorate was considered for regulation under the Stage 1 D/DBP Rule, but due to what the U.S. EPA considered a lack of sufficient health effects studies, chlorate was not regulated. As a result, the U.S. EPA asked the National Toxicology Program to evaluate the health effects of chlorate. The U.S. EPA may regulate chlorate in the future, although given the recently completed Stage 2 negotiations, there are no federal plans to do so without the additional health effects information.

Since the Stage 1 D/DBP Rule did not include chlorate, in January 2002, DHS reaffirmed the chlorate action level at 0.2 mg/L. The action level was based on a new evaluation of chlorate by the OEHHA. However, it was not DHS's intent to prohibit the use of ClO_2 as a disinfectant in California, rather, the AL was reaffirmed in response to groundwater that was contaminated with chlorate. In March 2002, DHS reconsidered chlorate as a disinfection byproduct and revised the action level to 0.8 mg/L to balance disinfection needs with health concerns.

Perchlorate

Perchlorate originates as a contaminant in the environment from the solid salt of ammonium, potassium or sodium perchlorate. Sources for the contamination include various chemical, industrial and agricultural uses. However, the major source of contamination appears to be the manufacture of ammonium perchlorate for the use as solid rocket fuel oxidizer in rocket motors, munitions, missiles and fireworks. The ammonium perchlorate provides the oxygen needed for the solid fuel to burn.

Prior to 1997, analytical detection limit for perchlorate was 400 ppb. However, due to concern over the presence of perchlorate in drinking water supply, the analytical method was refined and the detection limit improved to 4 ppb. The lowered detection capability has led to the discovery of perchlorate at various manufacturing sites and some drinking water supplies. In this early stage of occurrence data gathering, perchlorate has been detected in at least 14 states.

The USEPA has issued a draft risk assessment on perchlorate. The risk assessment concludes that perchlorate disrupts iodine uptake in the thyroid gland and poses risk for cancer and fetal development. Sensitive populations, like pregnant women, children, and people who have health problems or compromised thyroid conditions, should follow the advice of their health care provider regarding the amount and type of liquids, including water that should be consumed.

Until 1997, perchlorate was not considered a widespread contaminant, thus, drinking water standard was not established. Realizing the potential widespread problem for perchlorate and the probable adverse health affects, the EPA added perchlorate to the contaminant candidate list in March 1998. This opened the door for additional research in the area of health effects, treatment technologies, analytical methods and more complete occurrence data gathering. Perchlorate was included in the Federal Unregulated Contaminant Monitoring Rule and beginning January 2001, all large water system must begin occurrence monitoring for perchlorate. However, there is still no primary drinking water standard for perchlorate. In California, perchlorate was discovered in the San Gabriel Basin groundwater in May of 1997. This discovery prompted the California Department of Health Services (CDHS) to establish an action level at 18 ppb. Due to more recent occurrence and health effect information, CHDS lowered the action level for perchlorate to 4 ppb. Arizona, similar to California, established a health guidance level for perchlorate at 14 ppb.

Chlorine Dioxide Generators and Overview of Chlorine Dioxide Chemistry

Chlorine dioxide exists as a volatile, energetic free radical and is quite reactive. It is explosive at 5.8 psi (40 kPa) or above atmospheric pressure at 6 psig (41 kPa). For that reason, it cannot be compressed or stored and must be generated on-site (Aieta and Berg, 1986). Solutions generated for water treatment are generally in the range of 0.1 g/L to 5.0 g/L. If the aqueous concentration exceeds10 g/L, an explosion may occur (Aieta and Berg, 1986).

Chlorine dioxide does not hydrolyze in water and, for that reason, remains as a highly soluble gas when the water temperature is greater than 11°C and the pH is between 2 and 10. Chlorine dioxide solutions are greenish-yellow and smell strongly of chlorine (Gordon and Rosenblatt, 1996). Aqueous solutions must be protected from light, as ClO_2 is subject to photolysis, which results in ClO_2 degradation to chlorate (Zika, 1985). Griese et al. (1992) found chlorate concentrations ranging from 0.36 to 0.97 mg/L when he exposed water treated with from 3.6 to 4.0 mg/L ClO_2 to fluorescent lights. Water treated with ClO_2 and kept in the dark contained no chlorate. The significance of these properties of ClO_2 for water treatment is that because of its volatility, ClO_2 should be applied with a minimum of agitation (e.g. in rapid mixers) and exposure to light.

Most drinking water treatment plants using ClO_2 generate it from sodium chlorite ($NaClO_2$), either the solid form (Gordon and Rosenblatt 1996) or 25 percent solutions (Aieta and Berg, 1986). Sodium chlorite is then reacted with either chlorine gas (Cl_2), hypochlorous acid (HOCl), or hydrochloric acid (HCl) in the following reactions (Aieta and Berg, 1986; Gordon et al., 1972):

$$2NaClO_2 + Cl_{2(g)} \rightarrow 2ClO_{2(g)} + 2NaCl \quad (2.1)$$

$$2NaClO_2 + HOCl \rightarrow NaCl + NaOH + 2ClO_2 \quad (2.2)$$

$$5NaClO_2 + 4HCl \rightarrow 4ClO_{2(g)} + 2H_2O + 5NaCl \quad (2.3)$$

An unstable intermediate, Cl_2O_2, is formed during the first two reactions if the reactants are in high concentrations. When chlorite concentrations are low, the intermediate decays to chlorate. Equation (2.1) describes the reactions when either liquid or solid $NaClO_2$ is used. Chlorate ion can also be produced during reactions (2.1) and (2.2), both when the initial reactant concentrations are low and when the chlorine (or hypochlorous acid) concentration is in excess. When either the initial reactant concentrations are high or the ClO_2^- concentration is in excess, the intermediate decays to ClO_2 (Gordon and Rosenblatt, 1996). During reaction (2.3), only 80 percent of the $NaClO_2$ is converted to ClO_2, and, therefore, this method is not popular.

Ideally, ClO_2 generators should produce solutions free of by-products such as chlorite, chlorate, and chlorine. Unwanted by-products can arise from contaminated feedstock (e.g. with chlorate), the generation, when the process is not properly controlled, and when excess chlorine is added (Gates, 1998). Extremely high-purity ClO_2 gas can be generated when humidified chlorine gas is passed through solid $NaClO_2$, which is a method that became commercially available in the mid-1990s. During this process, no unreacted $NaClO_2$ enters the system because the generated ClO_2 is in the gas phase, chlorate is not produced, and no excess chlorine is present. A chlorine-free ClO_2 product stream is assured by passing the gas sequentially through two or more drums of solid $NaClO_2$ (Gordon and Rosenblatt, 1996).

The main reaction product formed when ClO_2 reacts with reduced substances (e.g. organic matter, sulfide, and manganous ion) in water is chlorite (ClO_2^-). The reaction involves transfer of a single electron as shown by the following half-reaction (Gordon et al., 1972):

$$ClO_2 + e^- \rightarrow ClO_2^- \quad (2.4)$$

Chlorite ion is also an oxidant, but it reacts at a much slower rate than ClO_2 under conditions generally encountered in water treatment. Chlorite is reduced to chloride (Cl^-) in the following reaction (Aieta and Berg, 1986; Gordon et al., 1972; Masschelein, 1979):

$$ClO_2^- + 4H^+ + 4e^- \rightarrow Cl^- + 2H_2O \quad (2.5)$$

Typically, from 50 to 70 percent of the applied ClO_2 yields chlorite with the balance forming chloride and chlorate (Aieta and Berg 1986, Limoni et al. 1984, Rav-Acha 1984; Werdehoff and Singer 1987). Chlorite ion can be removed when in excess by the addition of either reduced iron salts or reduced sulfur compounds.

In basic solutions, ClO_2 disproportionates to form chlorite and chlorate ions (Aieta and Berg, 1986):

$$2ClO_2 + 2OH^- \rightarrow ClO_2^- + ClO_3^- + H_2O \quad (2.6)$$

When chlorine is present, it reacts with ClO_2 to form chlorate (Aieta and Berg, 1986):

$$2ClO_2 + HOCl + H_2O \rightarrow 2ClO_3^- + 2H^+ + HCl \quad (2.7)$$

Utilities that apply ClO_2 as a preoxidant and add free chlorine for the distribution system residual often receive numerous customer complaints of odors variously described as "cat urine," "chlorine," "petroleum," and "kerosene-like" in customers' homes (Hoehn et al., 1990). Hoehn and his coworkers (1990) discovered the cause of these odors to be reactions between volatile organics in the household air and ClO_2 released from the water when a tap was opened in the home. In the distribution system, chlorite will react to form ClO_2 in the same manner it is generated at the treatment plant. New carpeting is a common source of the volatile organic compounds that react with ClO_2 in the air, but certain paints, hand soaps containing hydrocarbons, and other sources of volatile compounds have been identified. The authors of this study cited many reports linking ClO_2-related odor complaints to new carpeting. The odors can be avoided either by substituting chloramine for chlorine in the distribution system or by removing chlorite at the treatment plant. Contra Costa Water District will not face this problem because they ammoniate following chlorination and, therefore, eliminate one of the reactants (chlorine) that is responsible for regeneration of ClO_2.

Energy Demand Requirements for Water Treatment

The difference in energy required to generate ClO_2 versus ozone generation can be significant. According to data developed by Sterling Chemicals, one of the largest manufacturers of ClO_2 in the world, a conventional ClO_2 generator (like the one used in the initial studies at the treatment plant) uses 0.90 kWh per pound of ClO_2 generated at the treatment plant. However, as the state agency charged with assuring California's electric supply remains reliable, the California Energy Commission must consider the impact to the state's entire electrical system. With more than 2,500 million gallons of water treated with ozone per day in the state, the potential reduction in electricity consumption was immense. Thus, data generated in this study was used to estimate the total statewide demand reduction that could be achieved by substituting a portion of the ozone used in California water treatment plants with ClO_2.

Sterling has computed that the total electricity requirement for generating ClO_2 is 6.42 kWh per pound of ClO_2. Thus, for a treatment plant applying a dose of 1.5 mg/L of ClO_2, total on-site energy requirements would be about 11 kWh/MG water treated, while the total amount of electricity used to generate and apply ClO_2 is 78 kWh/MG.

By contrast, an air-fed ozone system typically consumes 100 to 400 kWh/MG water treated (EPRI 1999). Water treatment plants using ozone with liquid oxygen feed, such as the Randall-Bold plant, consume much less electric energy on-site (usually 50 to 100 kWh/MG) because the air preparation system accounts for an average of 40 percent of the total power requirement in a well-designed ozone system. However, measuring power consumption only at the treatment plant excludes the electric requirements associated with off-site generation of liquid oxygen. Typically, the electric needs are "included" in the delivered cost of the liquid oxygen. From the standpoint of the California Energy Commission, these electrical requirements must be considered because liquid oxygen is usually generated within the state, as is the case with the liquid oxygen purchased at the Randall-Bold plant.

A preliminary assessment, based on experiences at other water treatment plants using ozone in the United States, indicated that potential demand savings during peak energy use times at the Randall-Bold plant would approach 200 kW. However, if the same change was implemented at all water treatment plants in the state of California currently using ozone in their treatment scheme (treating an average of 2,575 mgd), the overall reduction in electric demand due to ClO_2 replacement in the state of California under average flow conditions was estimated to be nearly 6.9 MW. Much of these estimates were based on average or typical energy use factors associated with ozone generation. In reality, most of the average numbers are based on optimized systems and conditions which are not always encountered during processing, so the potential demand reduction could be even higher than the estimate. Both the actual reduction in ozone demand observed in this study, and the electric consumption data generated during the pilot trials, were used to verify this hypothesis.

DESCRIPTION OF RANDALL-BOLD WATER TREATMENT FACILITIES

Water Treatment Process Overview

The Randall-Bold WTP (Figure 2.1) is a direct filtration plant. It operates 24 hours per day, 7 days per week. As shown on Figures 2.1, the plant includes the following:

- Intake at the Contra Costa Canal. An approximately 160 feet long 48-inch main delivers water to the plant.
- Plant influent flowmeter and modulation facilities.
- Grit basin.
- Two 3-stage preozonation basins (countercurrent-cocurrent-countercurrent flow).
- Two hydraulic mixers.
- Two hydraulic flocculation basins.
- Six deep-bed GAC (80 inch) and sand (40 inch) filters equipped with surface wash and air scour and upflow backwash, with a surface area of 780 square feet each.
- Two 3-stage post-ozonation basins (countercurrent-cocurrent-countercurrent flow).
- Filter washwater basin, 250,000 gallons.
- Filter-to-waste basin, 250,000 gallons.
- Chlorine and ammonia feed facilities.
- A 5 MG, 22-ft deep (with sump) clearwell
- High-service pump station, four pumps.
- Chemical feed and storage facilities.
- Air dry solids lagoon with decant return.

Ozone is added to the raw water in the preozonation contactor to enhance filtered water quality and taste-and-odor control. The preozonation dosage typically ranges from 1.0 to 2.0 mg/L. Pathogen inactivation is achieved with post-ozonation, where the dose is typically 0.5 to 1.0 mg/L, and with chloramination in the clearwell. Pre-formed chloramines for distribution-system residual are formed prior to the clearwell by reaction of elemental chlorine in water solution and aqueous ammonia.

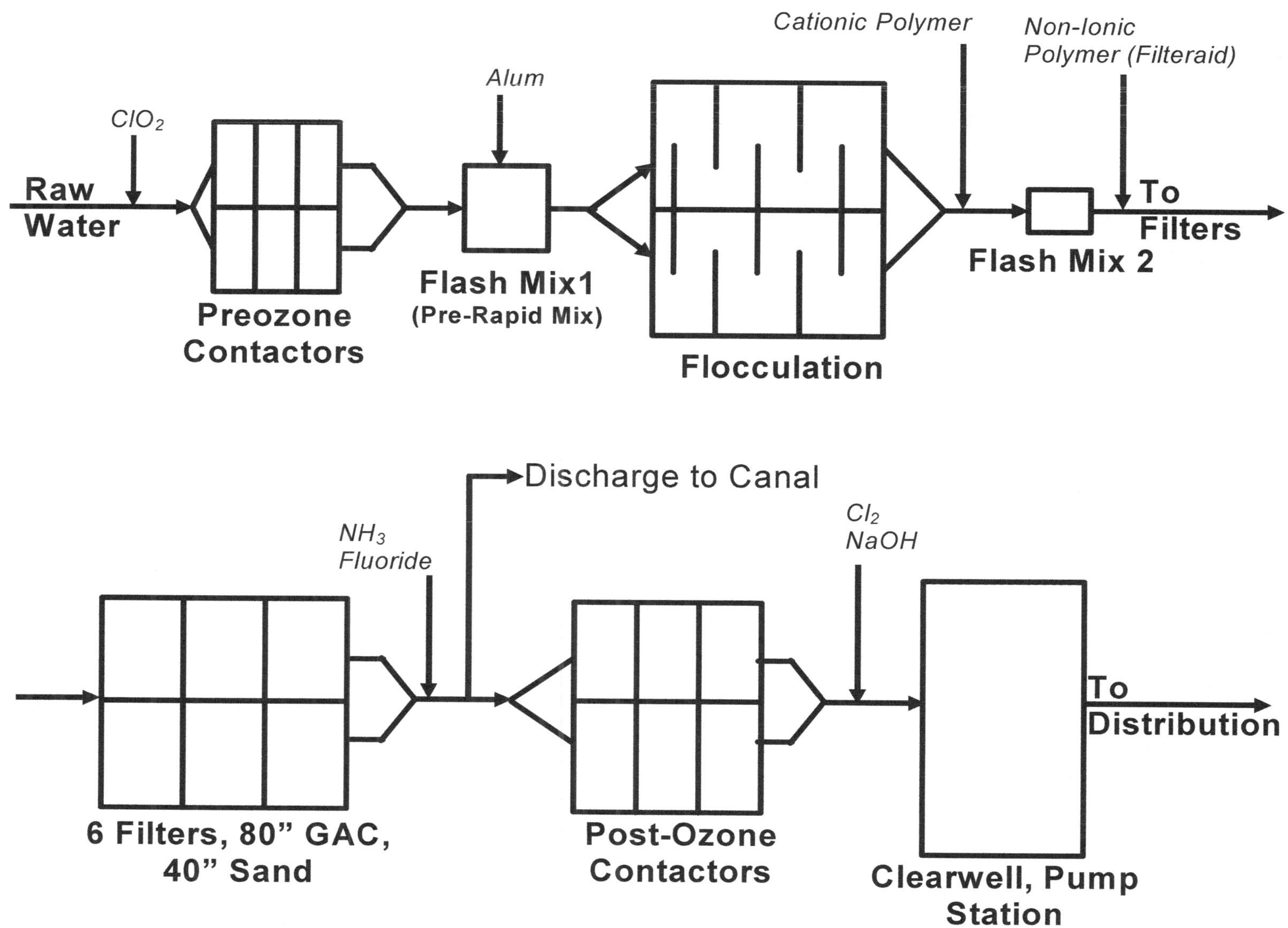

Figure 2.1 Schematic of Randall-Bold Water Treatment Plant

The hydraulic flocculation basins each contain 15 cells and can be operated at 20 to 30 mgd each. Temporary modifications have been made that closed the center of the cells to decrease detention time in the basins. The maximum design G value is approximately 70 sec^{-1}. Flocculation is provided by a system of adjustable gates and fixed baffles arranged to control turbulence in the basins. Energy is provided by the headloss across the basin, which can be maintained at approximately 1 foot of head by adjusting the gate positions over a range of flows.

The plant design capacity is 40 mgd, but, it is generally operated around 10 mgd. Last year, the plant was operated at 20 to 27 mgd in one of the two treatment trains during a study of high-rate filtration impacts on water quality and to verify the plant design capacity of 40 mgd. Half of the treated water was discharged to the Canal during this study. At relatively low flow rates (<5 mgd), the flocculation basins act as sedimentation basins and sludge accumulation poses problems. At design capacity, the filters will operate at 6 gpm/ft^2 but current operations are more in the range of 3-4 gpm/ft^2.

Filter Operating Policy

At the present time, only three filters of the six filters are in service for a flow of 20 mgd. Backwash water and filter-to-waste are stored in two separate 250,000 gallon lagoons. Attempts are made to keep the filter rates between 3 and 4 gpm/ft^2 on the three filters when the plant flow rate is 10 mgd. Filter backwash water is sent to the solids lagoon, then the decant from the solids lagoon are recycled to the influent.

HISTORICAL WATER QUALITY DATA

Raw Water

Raw water is provided through the Contra Costa Canal. Water in the Canal is derived from the Sacramento-San Joaquin Delta through various intakes (Figure 2.2). The two principal intakes are the Rock Slough Intake and the Old River Intake near Discovery Bay. Canal water is delivered to industrial users, public water supply retailers, and to CCWD's treatment facilities (Bollman WTP and the Randall-Bold WTP). Canal water derived from the Old River Intake can be stored in the Los Vaqueros Reservoir, which becomes the major source during periods when use of Delta water is prohibited. Water taken from the reservoir is replaced at relatively high expense incurred by pumping costs.

Raw-water quality at the Randall-Bold WTP is directly related to the use of either of the three raw-water sources or various combinations of all three. Important factors include but are not restricted to the following:

- Fish protection during "no diversion" periods
- Power-saving schemes
- Water quality in the Delta
- Canal operations and maintenance schedules
- Weather conditions and seasonal changes in rainfall
- Reservoir water level, refill schedule, and costs.
- Tidal influences from the San Francisco Bay

Tidal influences on water quality are especially important because they affect chloride and bromide ion concentrations in water delivered to the water treatment facilities.

The raw-water quality varies considerably during the year and is a function of the sources of water in the Canal. Table 2.2 summarizes the major raw water and finished water characteristics during 2000. Total organic carbon (TOC) concentrations (not shown in Table 2.2) in Randall-Bold WTP raw water during November 2, 2000 through March 28, 2001 averaged 4.43 mg/L. The high was 5.06 mg/L on March 14, 2001 and the low was 2.45 mg/L on November 2, 2000.

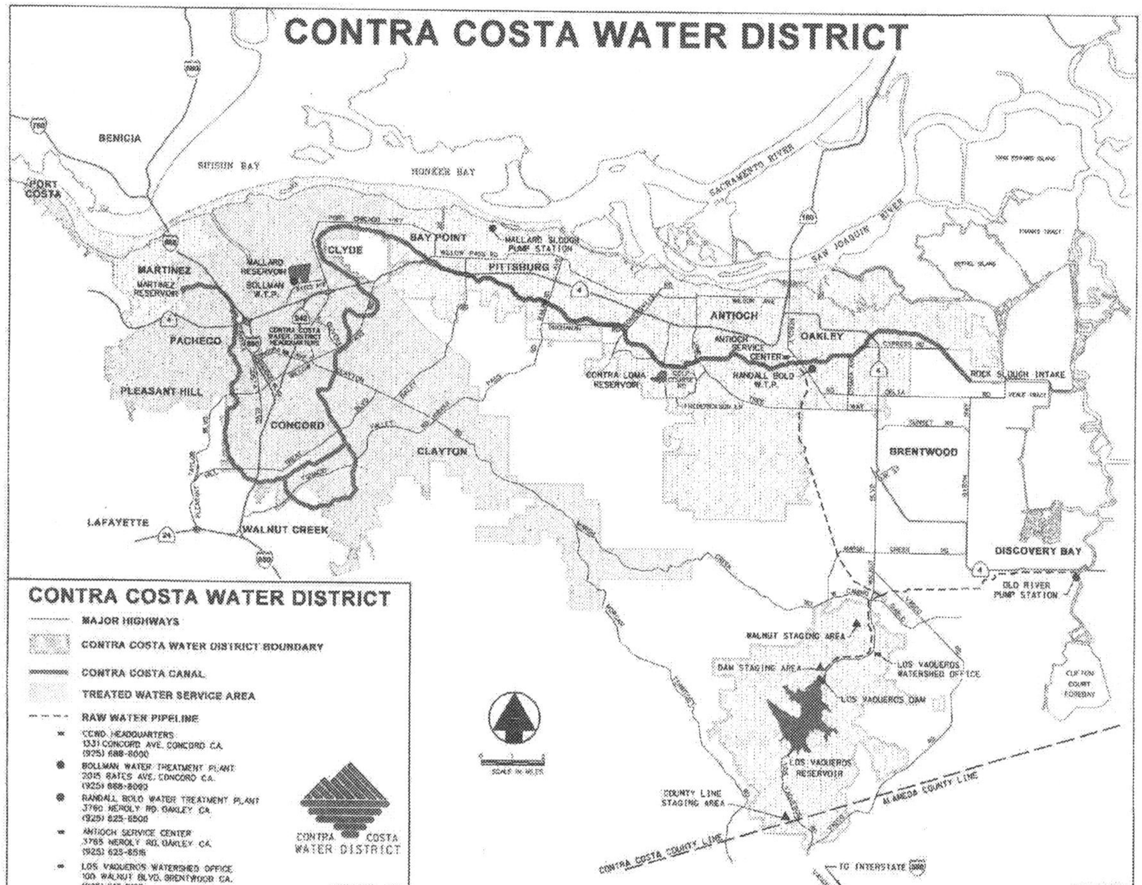

Figure 2.2 Contra Costa Water District source water map

Table 2.2
Historic water quality

	Raw Water				Finished Water	
Statistic	Turbidity (NTU)	pH	Alkalinity (mg/L as $CaCO_3$)	Bromide* (mg/L)	pH	Alkalinity (mg/L as $CaCO_3$)
Average	9.13	7.76	73.3	0.09	8.96	84.4
Median	9.44	7.73	67	0.14	8.94	79.5
Std. Dev.	4.13	0.21	12.8	0.09	0.22	12.5
High	18.77	8.25	98	0.26	9.98	106
Low	0.68	7.32	59	< 0.10	8.05	70
Number	41	41	41	28*	41	41

*Monthly samples Jan. 1998 – Sept. 2000; 13 samples < 0.1 mg/L. Statistics are for those > 0.1 mg/L

Finished Water

The major water-quality characteristics of finished water are shown in comparison to the raw-water quality characteristics in Table 2.2 above. Halogenated disinfection by-product (DBP) concentrations in finished water are quite low because the water is never in contact with free chlorine. As is shown in Figure 2.1, the ammonia is added ahead of the postozonation contactor and chlorine is not added until just prior to the clearwell. During 2000, monthly HAA concentrations in samples collected from five to six sites in the DWD system averaged <1.0 μg/L during April-August, 2000. In September and October, the averages were 8.1 μg/L and 5.1 μg/L, respectively. The total concentration was made up primarily of monochloroacetic acid, dichloroacetic acid, and trichloroacetic acid, with trichloroacetic acid accounting for most of the totals in each case. In November, only trichloroacetic acid was detected. Trihalomethane quarterly averages were always less than 5 μg/L throughout the system.

Bromate Formation

Monthly finished-water bromate concentrations have been collected at Randall-Bold WTP since early 1993, well before bromate compliance requirements. The MCL, which became effective January 1, 2002, is 10 ppb based on monthly sampling and a running annual average. A summary of the historic monthly bromate concentrations is shown on Figure 2.3 and the theoretical running annual average of the data since 1993 is shown on Figure 2.4.

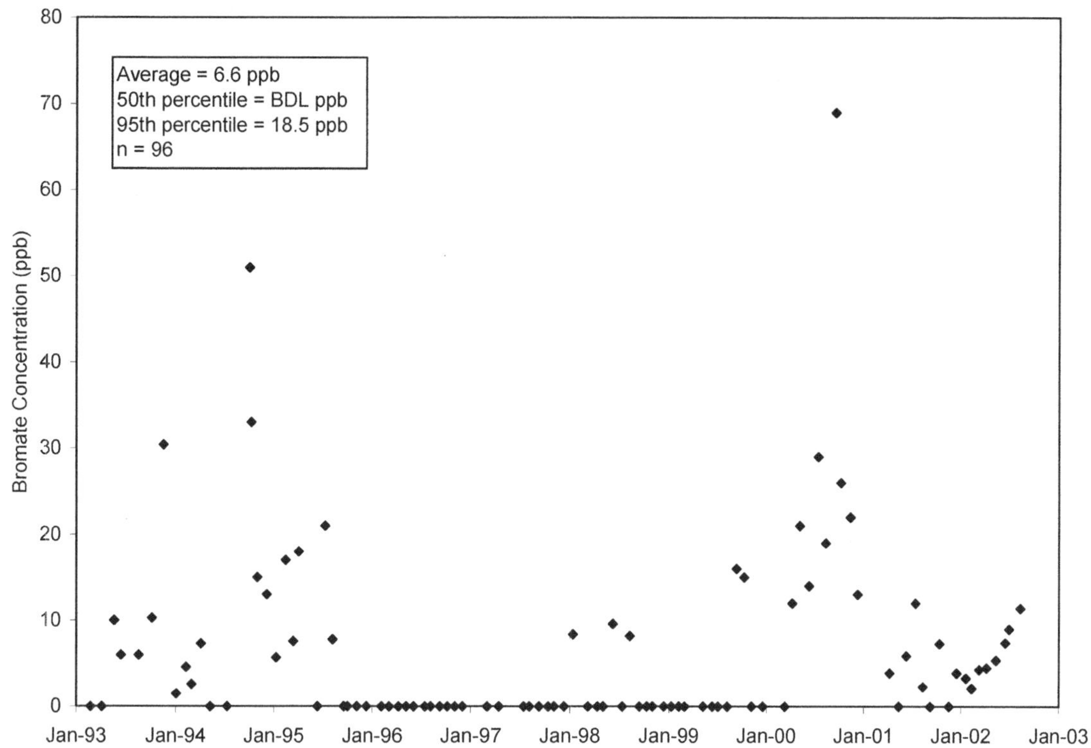

Figure 2.3 Historic monthly bromate concentrations for Randall-Bold WTP

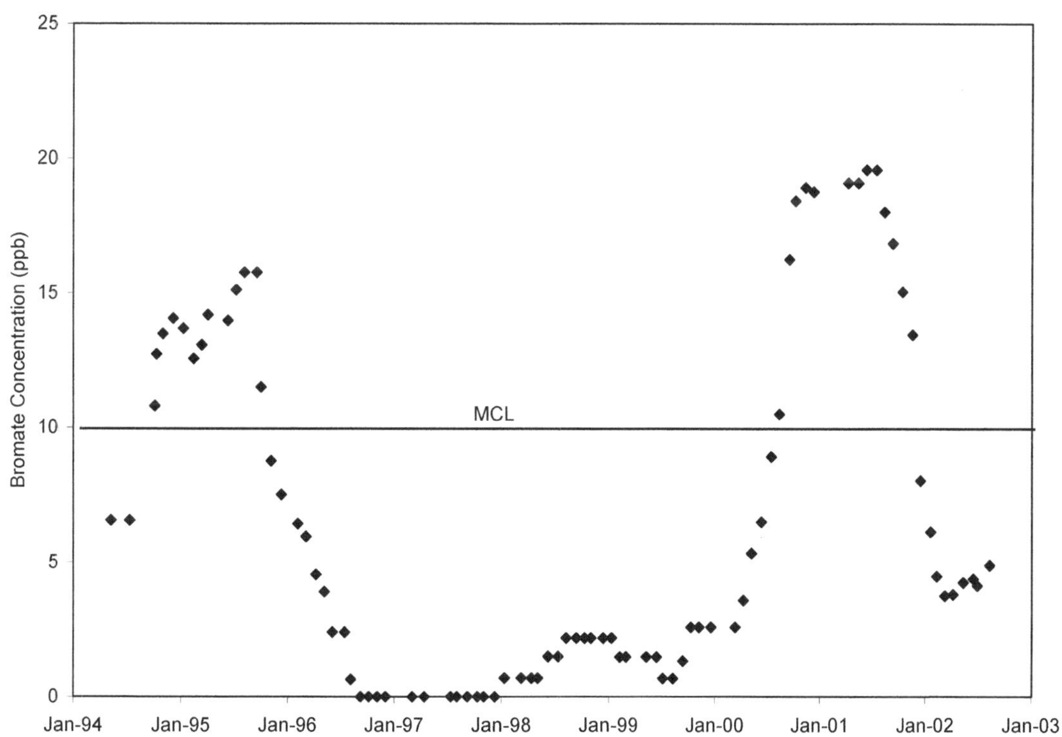

Figure 2.4 Running annual average bromate concentrations at Randall-Bold WTP

As shown in the previous figures, based on the historic data Randall-Bold could have problems complying with the current bromate MCL. Although the average bromate concentration of all data was 6.6 ppb and the concentration from 49 of the 96 samples was below the detection limit, the RAA shows that Randall-Bold would not have been in compliance for two different periods. The data also showed that the bromate concentration in 22 of the 96 samples was above the 10 ppb MCL. Therefore, unless the combination of ClO_2 and ozone lowers the bromate formation, the use of ozone at Randall-Bold may need to be reduced or even eliminated to comply with the bromate MCL.

CHAPTER 3
MATERIALS AND METHODS

ANALYTICAL METHODS

The analytical methods used during this project were those specified by the USEPA. The methods are used routinely at CCWD, and their laboratory is certified for compliance sampling and analysis. The CCWD Laboratory is state-certified to perform the analyses required for compliance with the D/DBP Rule. Assimilable organic carbon (AOC), N-nitrosodimethylamine (NDMA), perchlorate, and some bromate analyses were contracted to an outside laboratory (MWH Laboratory, Pasadena, California).

The sampling protocols and analytical methods are as follows:

- ClO_2/ClO_2^-: Standard Method 4500-E, amperometric titration for daily analyses of the generator effluent and water entering the distribution system (APHA et al. 1998). Process-control samples of ClO_2 were analyzed with the lissamine green B (LGB) colorimetric method or the "one-step method", which uses Step A from 4500-E as an estimate of the chlorine dioxide residual. Monthly compliance ClO_2^- analyses were by EPA Method 300.1, ion chromatography (IC). Samples for IC analyses of ClO_2^- were preserved at the time of collection with ethylene diamine, as specified in the method.
- ClO_3^-: Chlorate ion monitoring is not currently required by the D/DBP Rule, but it was determined in generator-effluent samples and finished-water samples by EPA's IC Method 300.1.
- TOC: Method 5310 (either B, C, or D) in *Standard Methods for the Examination of Water and Wastewater* (APHA et al. 1998) was followed. If the analyses were not performed immediately, the sample pH was adjusted to pH 2 with sulfuric acid and refrigerated. The recommended quality-control samples were also analyzed.
- SUVA: Specific ultraviolet absorbance (SUVA) analyses was determined in addition to TOC, even though it is specified as an alternative criterion for demonstrating compliance with the TOC-removal requirements. The sample absorbance at 254 nm and TOC concentration are required for calculations of SUVA. Absorbance was determined according to Standard Method 5910B, 1998 edition.
- Br^-: Bromide ion was determined by ion chromatography, EPA Method 300.1.
- BrO_3: Bromate ion was determined by ion chromatography, EPA Method 300.1. A majority of the samples were analyzed by the CCWD Laboratory, which had a detection limit of 1 or 2 ppb, depending upon standard calibration curves. Some samples were sent to an outside laboratory that had a detection limit of 5 ppb.
- Chlorine residuals were determined by amperometric titration according to Standard Methods 4500-Cl D and E (APHA 1998).
- Simulated Distribution System (SDS) Test: This procedure for determining THM and HAA formation potentials was used during this project. This test is similar to the Uniform Formation Conditions (UFC) test described by Summers et al. (1994) except that the samples are incubated for a period that mimics the maximum residence time in the utility distribution

system, in this case 3 days. The samples were buffered at pH 8 and the chlorine concentration increased to a level that will provide a residual of 1.0 $\pm$ 0.3 mg/L at the end of the incubation period. The incubation temperature was 20°C. The THM and HAA formation potentials were determined according to methods noted below.

- THMs: the four THMs currently included in the D/DBP Rule ($CHCl_3$, $CHBrCl_2$, $CHBr_2$, and $CHBr_3$) were determined by EPA Method 524.2, which is used for determination of Volatile Organic Chemicals.
- HAAs: The five HAAs currently included in the D/DBP Rule (mono-, di-, and trichloroacetic acids and mono- and dibromoacetic acids) were analyzed by EPA Method 552.

SAMPLING LOCATIONS

The sampling locations during the testing were as follows:

- Raw Water – The raw water was sampled on at least a weekly basis to determine raw-water quality and to quantify any variations with time.
- XI-106 – This sample location was located about 62 feet downstream of the ClO_2 feed point on the raw water pipeline and was used to measure the initial ClO_2 residual.
- Post ClO_2 – Samples of raw water that have been treated with ClO_2, but not ozone, were collected from a sample line located at the end of cell 1 in the preozonation contactor.
- Pre-Rapid Mix – Samples of water leaving the preozonation contactor were collected at the rapid mixer because a sampling point is not available at the effluent end of the preozonation contactor.
- Filter Effluent – Filter effluent samples were collected at the sample tap located in the 54-inch pipe at the inlet to the post-ozonation basin.
- Postozonation Effluent – Samples following post ozonation were collected from the effluent of the contactor.
- Plant Effluent – The plant effluent was sampled from the sampling tap located inside the clearwell on the 30-inch clearwell-effluent line.

CHLORINE DIOXIDE GENERATION SYSTEM

During the study, two ClO_2 generation systems were used to conduct the testing. The first system was a two-chemical system that was used to conduct all of the testing in 2001 and in January 2002. The second system was a single chemical electrochemical system that was used during July 2002. Both systems were flow paced to ensure the target dosage was delivered as flow changed. The electrochemical system was tested because it can generate a higher purity ClO_2 and it required less daily operator attention. Although an electrochemical generator was used for the July 2002 test, the results should be comparable with testing done using the two-chemical generator. The following section provides a description of each system.

Electrochemical System

The electrochemical system used was an ERCO R101™ Electrochemical generator rated at 150 pounds of ClO_2 per day, from Sterling Pulp Chemicals Inc., Water Technologies Division.

This system generates a ClO_2 solution from sodium chlorite and water using an electrochemical cell and hydrophobic membrane. The chlorite is converted to ClO_2 in the anode portion of the cell, which is then separated from the anolyte flow via a hydrophobic membrane. This membrane allows only ClO_2 as a gas and water vapor to transfer into an absorbent solution. The absorbent solution, in this case potable water, on the other side of the membrane, receives the ClO_2 and water vapor. This produces essentially pure ClO_2 solution that is then fed to the application point using the batch tank and feed control system as used in the two-chemical system. A schematic of the electrochemical system is shown on Figure 3.1.

Two-Chemical System

The two-chemical system was an Oxychlor II, manual ClO_2 generator rated at 150 pounds per day, from International Dioxcide, Inc. (IDI). The generator used the reaction between liquid sodium chlorite and chlorine gas to generate a ClO_2 solution. The liquid sodium chlorite and chlorine gas were mixed in the water phase by an eductor and allowed to react in a reaction column. The chemical feed rates were controlled automatically and set based on the plant flow rate and desired ClO_2 dose. Adjusting the chlorine gas feed rate to maintain a target pH range of 3.8 to 4.2 in the ClO_2 solution minimized excess chlorine. The ClO_2 solution was sent from the generator to a batch tank. Measuring the concentration of the ClO_2 solution on-line and integrating this with the required dose rate and plant water flow, the solution was then pumped via a control valve arrangement to the chemical application point at the head of the plant. A schematic of the generator system and the ClO_2 feed points are shown on Figure 3.2.

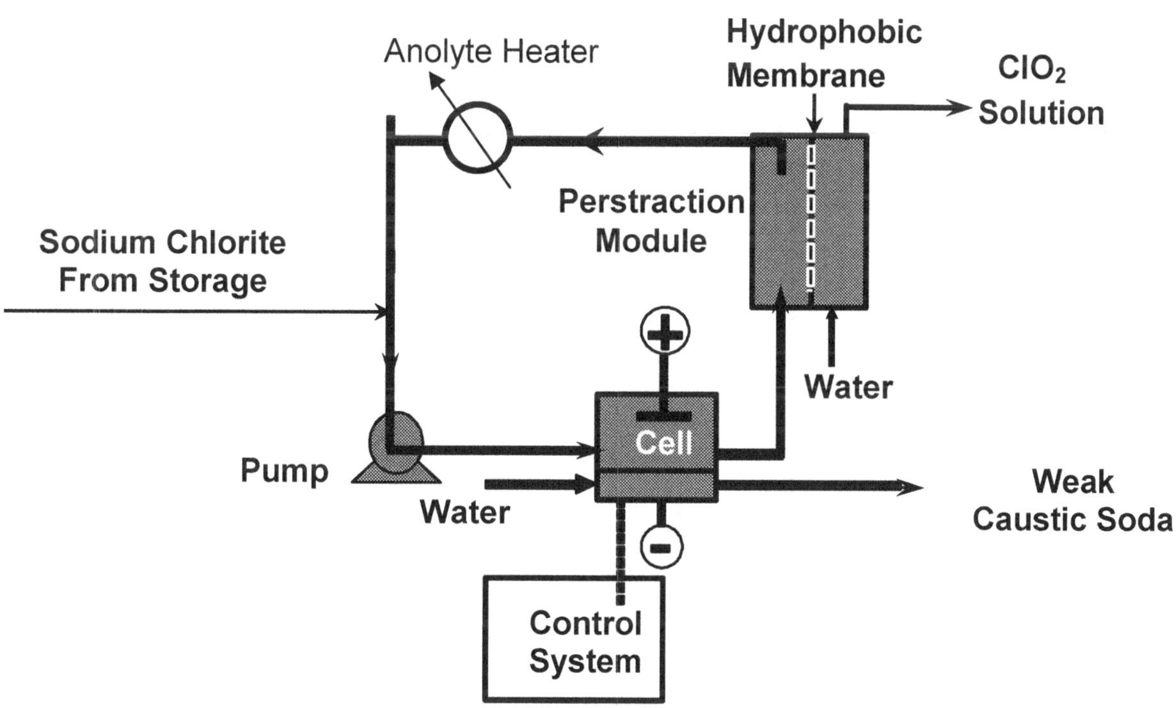

Source: ERCO Worldwide
Figure 3.1 Schematic of the ERCO R101™ ClO_2 generator

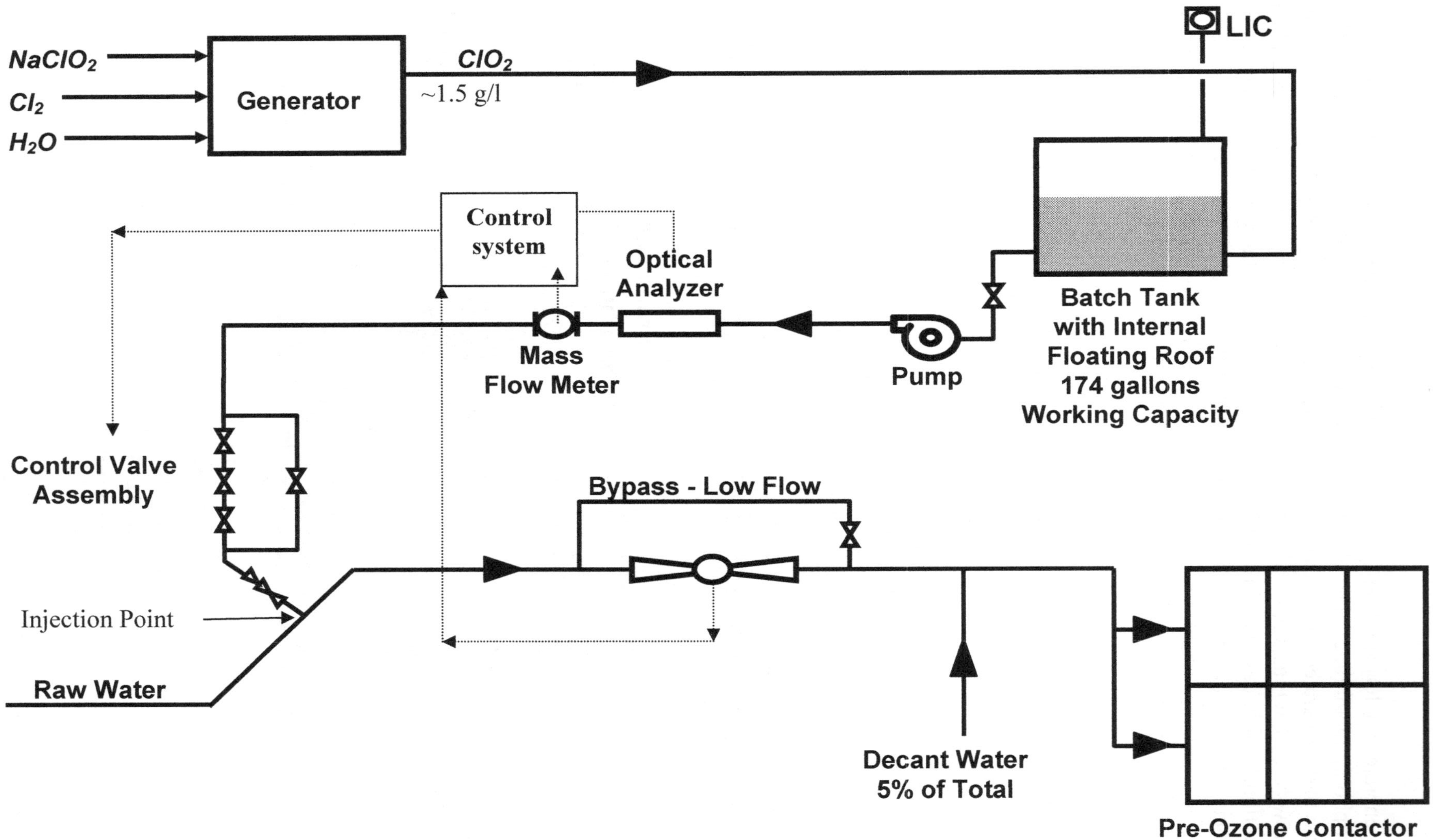

Figure 3.2 ClO_2 feed system schematic

OZONE GENERATION SYSTEM

The ozone generation system at the Randall-Bold WTP was manufactured by Ozonia North America and has been in service since 1992. The system consists of 3-450 pound per day (ppd) ozone generators that normally generate 7 to 10 percent weight ozone from oxygen. The liquid oxygen (LOX) is stored in a 6,000 gallon tank and four ambient vaporizers to convert the LOX to gaseous oxygen (GOX). Ozone dosages are flow paced and ozone can be added to each cell in the ozone contactors using rod diffusers. The plant is typically operated with one generator for the preozonation contactor and one generator for the postozonation contactor.

ACID FEED SYSTEM

A temporary acid feed system was used during the study. The system consisted of a 1,000 gallon storage tank and a Milton Roy metering pump. The system was located adjacent to the raw water piping and acid was supplied just upstream of the decant return line, which is before the preozonation contactor. Sulfuric acid was used for the pH adjustment tests.

AMMONIA FEED SYSTEM

The existing Randall-Bold ammonia system, which uses 19 percent aqua ammonia, was used for the ammonia mitigation testing. Piping already existed from the ammonia feed room to the raw water, so only slight modifications to the metering pump piping were required.

ENERGY CONSUMPTION MEASUREMENTS

Energy consumption data was generated using a combination of measured and calculated information. The electricity used at the treatment plant to generate both ClO_2 and ozone is easily measured by monitoring the respective generators. However, as the energy analysis addressed the implications of using ClO_2 for the entire state's electricity demand, the electricity needed to synthesize ClO_2's raw materials must also be considered.

As noted previously, ClO_2 is synthesized electrochemically from chlorine and sodium chlorite, both of which are delivered to the plant. On the other hand, sodium chlorite is made commercially from sodium chlorate, hydrochloric acid, hydrogen peroxide, and sodium hydroxide. The production of all of these constituents requires modest amounts of electricity, but the majority of the need comes from sodium chlorate. This compound is generated in an electrochemical cell from sodium chloride, caustic, hydrochloric acid and hydrogen peroxide. The electrochemical reaction accounts for 96 percent of the electricity needed to synthesize the sodium chlorite. Based on this chemistry, it was computed that 6.43 kWh of electricity is needed to generate one kilogram of ClO_2.

As with ClO_2, ozone generation entails significant electricity use to produce the raw materials. Each ozone generator at the Randall-Bold plant is equipped with a watt meter to record energy consumption. This information is collected and recorded daily. The electricity consumption up to one year prior and during the pilot study was analyzed to determine on-site energy use. In addition, liquid oxygen (the feed gas for ozone at the treatment plant) is prepared through a cryogenic process, where air is cooled and distilled to separate nitrogen and oxygen. In a simplified process, air is compressed to about 90 psia, cooled, and the liquefied air is distilled under high pressure. Cryogenic production of oxygen requires 260 to 340 kWh per ton of oxygen produced, depending on the size of the cryogenic plant. Very small plants (i.e. producing less than 20 tons of oxygen per day) require higher amounts of electricity. For purposes of this analysis, an average value of 300 kWh per ton of oxygen was assumed.

CHAPTER 4
TESTING PROTOCOL

INTRODUCTION

The testing during the project was divided into three phases that acted as building blocks to take the evaluation from bench-scale ClO_2 demand studies to long-term full-scale operation using ClO_2 in combination with ozone. The first phase of testing consisted of screening different dosages of ClO_2 and ozone combinations for further testing. The screening involved four-hour tests, with samples collected only at the preozonation contactor for a quick evaluation of viable dosage combinations based on the ozone demand/dose, and bromate and chlorate formation results. During the second phase of testing, selected dosage combinations of ClO_2 and ozone were evaluated by short-term tests of two days' duration, which is longer than the typical filter run of 24 hours. These tests were designed to evaluate the four research objectives as well as the effect of substituting ClO_2 for ozone on filter effluent quality. The third and final phase consisted of three-week "long-term" tests to evaluate the four research objectives as well as the operational challenges that might be encountered when using a combination of ClO_2 and ozone. Three preoxidation conditions were tested: 1.5 mg/L of ClO_2 only, the combination of 1.0 mg/L of ClO_2 and 0.5 mg/L of ozone, and 1.0 mg/L of ozone only. Each condition was evaluated for three weeks, with one replicate, to provide 18 weeks of full-scale operation.

When ClO_2 was applied during this project, cell 1 of the three-cell preozonation contactor was used to provide additional ClO_2 contact time and ozone was added in cell 2 only for testing both with and without ClO_2. The target ClO_2 dosage was low enough to leave no residual as the water entered the second cell of the contactor, where ozone was added. The research plan incorporated safeguards to ensure that partial elimination of ozonation in the preozonation contact chamber did not compromise treatment, and a water that did not meet the quality standards was discharged to the Contra Costa Canal.

PHASE I - SCREENING TESTS

Selected combinations of ClO_2 and ozone dosages were evaluated by conducting 4-hour tests that focused primarily on the preozonation contactors. Ozone demand and decay, bromate formation, and chlorate formation data were used to determine the impacts of varying the ozone dose while the ClO_2 dose remained constant, and varying the ClO_2 dose while the ozone dose remained constant. Testing was conducted over a one-week period, and the one-week screening test was conducted twice at the start of the testing program to provide replicate data. After the initial screening tests, combinations of ClO_2 and ozone dosages were selected for further testing during the short-term testing phase.

Testing Plan

The week-long screening tests evaluated three different ozone doses and four different ClO_2 doses. The ozone dose was reduced from the typical plant dose of 1.5 mg/L to 1.0 and 0.5 mg/L. The three ClO_2 doses to be used during that particular week were determined by bench-scale demand studies conducted at the beginning of each week. Control tests (i.e., no applied ClO_2, but with ozone added) were conducted. The three ClO_2 doses were in the range from zero to the maximum dose (likely 1.0 mg/L). Table 4.1 provides an example of the weekly plan that was executed during the screening tests.

Table 4.1
Screening tests - weekly plan

Day	Test start time (hours)	Test end time (hours)	O_3 dose (mg/L)	ClO_2 dose* (mg/L)
Monday	-	Day shift	-	Conduct bench demand study
	1330	Overnight	1.0	0.0
Tuesday	Overnight	0530		
	0530	0930	1.5	0.0
	0930	1330	0.5	0.0
	1330	Overnight	1.0	0.25
Wednesday	Overnight	0530		
	0530	0930	1.5	0.25
	0930	1330	0.5	0.25
	1330	Overnight	1.0	0.5
Thursday	Overnight	0530		
	0530	0930	1.5	0.5
	0930	1330	0.5	0.5
	1330	Overnight	1.0	1.0
Friday	Overnight	0530		
	0530	0930	1.5	1.0
	0930	1330	0.5	1.0
	1330	-	-	-

*Actual ClO_2 dosages were determined after results of bench scale demand study

As shown in Table 4.1, Monday was dedicated to the bench-scale ClO_2 demand study and initiating the baseline testing condition (ozone dosage of 1.0 mg/L and no ClO_2). During each subsequent day, three different 4-hour tests were conducted. The plant was operated under one set of conditions overnight, until 0530 hours, then conditions were changed and the plant was operated under the new set of conditions until 0930 hours. Conditions were changed again at 0930 hours, and once again at 1330 hours. Plant operation was continued overnight under the latter set of conditions. Because experience has shown that ozonation is important for maintaining filtered water quality, the ozone dose for overnight conditions was the same as that typically applied in the preozonation chamber (1.5 mg/L). Even though the experimental plan called for increasing the ClO_2 dose each day, ozone was added at a high enough dose to ensure the desired residual in the preozonation chamber effluent. Furthermore, using a high enough ozone dose overnight to maintain the desired filtered water quality should have ensured that ClO_2 addition during the day did not result in degraded finished water quality overnight. The actual ClO_2 and ozone doses used during weeks 1 and 2 of the screening test are listed in Table 4.2.

Table 4.2
Testing matrix for screening tests

Run	Week 1 ClO_2 dose (mg/L)	Week 2 ClO_2 dose (mg/L)	Ozone dose (mg/L)
1	-	-	1.0
2	-	-	1.5
3	-	-	0.5
4	0.25	0.5	1.0
5	0.25	0.5	1.5
6	0.25	0.5	0.5
7	0.50	0.75	1.0
8	0.50	0.75	1.5
9	0.50	0.75	0.5
10	0.75	1.0	1.0
11	0.75	1.0	1.5
12	0.75	1.0	0.5

Sampling Schedule

Sampling during each set of test conditions coincided with the operator's daily sampling schedule at 0500, 0900, and 1300 hours. Sampling, therefore, was always done at the end of each 4-hour evaluation of a given set of experimental conditions. Table 4.3 lists the sampling locations and analyses during the screening tests. Weekly samples were collected Wednesday and daily samples were usually collected during the 1300-hour sampling event.

PHASE II – SHORT-TERM TESTS

Phase II of the testing focused on short-term, full-scale testing of ClO_2 and ozone combinations to generate data for evaluation of all four testing objectives. Testing was conducted over a period of at least 48-hours so that several filters could be operated under a given set of test conditions until they needed backwashing. The test conditions included ozone alone and ozone in combination with ClO_2 at dosages based on the results of the screening tests. Six different combinations of dosages were selected for further evaluation.

Table 4.3
Sampling schedule - screening tests

	Number of samples per 4-hour test					
	Raw water	Post-ClO_2	Pre-rapid mix	Filter effluent	Post-O_3 effluent	Plant effluent
Turbidity	F					
pH	F					
Hardness	F					
Alkalinity	F					
Ammonia	F					
Temperature	F					
Iron	W			W		
Manganese	W			W		
TOC/DOC	W		W	W	W	W
UV254	W	W	W	W	W	W
Bromide	D					
Bromate	W		1			
Chlorite	W	1	1			
Chlorate	W	1	1			

F – Full-scale data
W – Weekly samples
D – Daily samples

Testing Plan

The short-term testing was divided into three 48-hour tests per week. Each test series included an evaluation of ozone alone and ozone in combination with ClO_2 at two different dosages. Evaluations of ozone without ClO_2 were conducted over the weekend, and the two oxidants in combination were evaluated during the week. The evaluation of ozone alone began on Friday afternoon and continued until Monday morning. Evaluation of ozone in combination with ClO_2 at one dosage began immediately thereafter and continued until Wednesday morning. Next, both the ClO_2 and the ozone dosages were changed, and the second oxidant combination was evaluated during the next 48 hours, terminating on Friday morning. At that time, the ClO_2 feed was turned off, and only ozone was added during the weekend. On the following Monday, the tests on combinations of ozone and ClO_2 were resumed.

Full-scale operating data were collected from the plant SCADA system to compare the results of the short-term tests. Filtered water turbidity was logged every 15 minutes to an Excel spreadsheet. The filter run time, headloss, and applied turbidity were also monitored from plant data. Information on the ozone system was collected to monitor the ozone dose and ozone residual in the preozonation and postozonation contactors during each testing condition.

A summary of the test start and stop dates and the ClO_2 and ozone doses for the short-term tests is presented in Table 4.4.

Table 4.4
Testing matrix for short-term tests

Test number	Start date	End date	ClO_2 dose (mg/L)	O_3 dose (mg/L)
1	7/27 (F)	7/30 (M)	-	1.0
2	7/30 (M)	8/1 (W)	-	-
3	8/1 (W)	8/3 (F)	1.0	-
4	8/3 (F)	8/6 (M)	-	1.0
5	8/6 (M)	8/8 (W)	1.0	1.0
6	8/8 (W)	8/10 (F)	0.5	1.5
7	8/10 (F)	8/13 (M)	-	1.5
8	8/13 (M)	8/15 (W)	0.5	1.0
9	8/15 (W)	8/17 (F)	1.0	1.0
10	8/17 (F)	8/20 (M)	-	1.0
11	8/21 (T)	8/24 (F)	0.5	1.0
12	8/24 (F)	8/27 (M)	-	1.0

Sampling Schedule

The sampling schedule for the short-term testing is listed in Table 4.5. Samples analyzed for evaluation of each test condition were collected near the end of the 48-hour run, at 0800 hours on Monday, Wednesday, and Friday. Weekly samples were collected on Monday at 0800 hours.

PHASE III – LONG-TERM TESTS

The final testing phase, Phase III, focused on the long-term, full-scale operation of the Randall-Bold WTP with combinations of ClO_2 and ozone. Selected combinations of ClO_2 and ozone, ClO_2 alone, and ozone alone were evaluated further during three-week long test runs. During the final four months of the testing program, ozone alone was evaluated on two separate occasions, and each of the selected oxidant combinations was evaluated on two separate occasions.

Table 4.5
Sampling schedule - short-term tests

	Number of samples per 48-hour test					
	Raw water	Post ClO_2	Pre-rapid mix	Filter effluent	Post O_3 effluent	Plant effluent
Turbidity	F					
pH	F					
Hardness	F					
Alkalinity	F					
Ammonia	F					
Temperature	F					
Iron	1			1		
Manganese	1			1		
TOC/DOC	1		W	W	W	W
UV254	1	W	W	W	W	W
Bromide	1					
Bromate	W		1		1	
Chlorite	W	1	1	1	1	1
Chlorate	W	1	1	1	1	1
THM						1*
HAA						1*
UFCTHM	1*			1*		
UFCHAA	1*			1*		

F – Full-scale data.
W – Weekly samples.
* Collected once per week, test conditions varied.

Testing Plan

Full-scale operating data were collected from the plant SCADA system to compare the test conditions. Filtered water turbidity was logged every 15 minutes to an Excel spreadsheet. The filter run time, headloss, particle counts, and applied turbidity were also monitored from plant data. Information on the ozone system was collected to monitor the ozone dose and ozone residual in the preozonation and postozonation contactors during each testing condition. The testing phase, testing week, start date, ClO_2 dose, and ozone dose for the long-term testing are listed in Table 4.6.

Table 4.6
Long-term testing plan

Phase	Week	Start Date	ClO_2 Dose (mg/L)	O_3 Dose (mg/L)
3A	1	8/27	1.5	-
	2	9/3	1.5	-
	3	9/10	1.5	-
3B	4	9/17	1.0	0.5
	5	9/24	1.0	0.5
	6	10/1	1.0	0.5
	7	10/8	-	1.0*
3C	8	10/15	-	1.0
	9	10/22	-	1.0
	10	10/29	-	1.0
3D	11	11/5	1.5	-
	12	11/12	1.5	-
	13	11/19	1.5	-
3E	14	11/26	1.0	0.5
	15	12/3	1.0	0.5
	16	12/10	1.0	0.5
3F	17	12/17	-	1.0
	18	12/24	-	1.0
	19	12/31	-	1.0
	20	1/7	-	1.0
	21	1/14	Screening	Screening

* During high-rate testing

Sampling Schedule

The sampling frequency for the long-term testing is listed in Table 4.7. Samples of constituents tested once per week were collected on Tuesday. Samples of constituents tested twice per week were collected on Tuesday and Friday.

BROMATE MITIGATION TECHNIQUE COMPARISON

The expanded objectives of the project included evaluation of bromate mitigation techniques for potential future application at the Randall-Bold WTP. These techniques included preoxidation with ClO_2, pH adjustment prior to preozonation, and ammonia addition prior to preozonation. The evaluations were conducted under conditions similar to those established for the screening tests described previously, with the exception that the three bromate-mitigation treatments were substituted for variations in the ozone and ClO_2 dosages. The mitigation techniques were evaluated during 2-hour tests that focused on the preozonation contactor and were conducted over a period of several days. The testing was conducted twice during the study, once during January 2002 just after the Phase 3 testing and once during testing in July 2002.

Table 4.7
Sampling schedule - long-term tests

	Number of samples per week					
	Raw water	Post-ClO_2	Pre-rapid mix	Filter effluent	Post-O_3 effluent	Plant effluent
Turbidity	F					
pH	F					
Hardness	F					
Alkalinity	F					
Ammonia	F					
Temperature	F					
Iron	1			1		
Manganese	1			1		
TOC/DOC	1		1	1	1	1
UV254	2	1	1	1	1	1
Bromide	2					
Bromate	1	2	2	2	2	2
Chlorite	1	2	2	2	2	2
Chlorate	1	2	2	2	2	2
THM						1
HAA						1
UFCTHM	1			1		
UFCHAA	1			1		
AOC	C		C	C		C
NDMA						

F – Full-Scale data
C – Once per testing condition

During January, tests were conducted for bromate mitigation and special tests were conducted to determine the impact of ClO_2 preoxidation on ozone demand and decay. The bromate mitigation tests included ozone only, ClO_2 preoxidation, pH adjustment, and ammonia addition. During tests with ozone only, ozone was added in cell 2 at a dosage of 2.0 mg/L. Preoxidation with ClO_2 was tested at dosages of 0.5 and 1.0 mg/L. Adjustment of pH was evaluated at pH values of 6.5 and 6.0 and ammonia was added at 0.25 and 0.5 mg/L NH_3-N. Each test lasted about 2 hours, which was equivalent to 4 hydraulic retention times. Sampling occurred at the end of the 2-hour tests and was done in duplicate. Each of the conditions was also repeated for a total of 14 tests. Analytical samples for the test were collected according to the schedule in Table 4.8.

A special series of four 2-hour tests were also conducted to determine the impact of ClO_2 on the ozone demand and decay. The first test consisted of ozone only at a dosage of 3.6 mg/L, which was the dose required to produce measurable residuals at the end of cell 2 and 3. For the remaining three tests, the ozone dose was held constant and ClO_2 was added at dosages of 0.25, 0.5, and 1.0 mg/L. Near the end of the tests, bromate, chlorite, and chlorate samples were collected and the ozone residual at the end of cell 2 and 3 was measured.

The bromate mitigation testing was repeated in July to verify and expand on the results from January. A total of twelve 2-hour tests were conducted to evaluate the bromate reduction when preoxidation with ClO_2, sodium chlorite, pH adjustment, or ammonia addition were used. Chlorine dioxide addition was evaluated at dosages of 0.5 and 1.0 mg/L and sodium chlorite was added at 0.7 mg/L to simulate the chlorite formation from the 1.0 mg/L ClO_2 dose. Adjustment of pH was evaluated at values of 6.5 and 6.0, and ammonia was added at 0.25 and 0.5 mg/L NH_3-N. Analytical samples for each test were collect at 100, 110, and 120 minutes into the test according to the schedule in Table 4.8.

Table 4.8
Sampling schedule – bromate mitigation tests

	Number of samples per 2-hour test					
	Raw water	Post-ClO_2	Pre-rapid mix	Filter effluent	Post-O_3 effluent	Plant effluent
Turbidity	F					
pH	F					
Hardness	F					
Alkalinity	F					
Ammonia	F					
Temperature	F					
Iron	W					
Manganese	W					
TOC/DOC	D					
UV254	D					
Bromide	D					
Bromate	W		2/3*			
Chlorite	W	2/3*	2/3*			
Chlorate	W	2/3*	2/3*			

F – Full-scale data
W – Weekly samples
D – Daily samples
*2 samples per test during January and 3 samples per test during July

CHAPTER 5
PHASE I – SCREENING TESTS

To test applicable combinations of ClO_2 and ozone dosages, a two-week series of 4-hour screening tests was completed during the weeks of July 9 and July 16, 2001. During both weeks, the ClO_2 doses ranged from zero to 1.0 mg/L, with ozone doses of 0.5, 1.0, and 1.5 mg/L. Detailed information on the testing plan and a summary of the ClO_2 and ozone doses used during each week is presented in Chapter 4. During the screening tests, the evaluation focused on the preozonation contactor, specifically, on the bromate, chlorate, and chlorite formation as well as on the ClO_2 and ozone residuals produced at various dosages. The following chapter describes the bromate formation, chlorite/chlorate formation, the ClO_2 and ozone demand, and the conclusions derived from an analysis of the data.

RAW WATER QUALITY

The raw water quality was monitored daily for various constituents based on the sampling schedule developed in the testing protocol. Raw water samples were normally collected around midday prior to any chemical addition. The raw water quality did not change substantially either from day to day or from week to week during the two-week period. The average raw water concentrations of alkalinity, ammonia, bromide, hardness, pH, temperature, turbidity, TOC, and UV254 are presented in Table 5.1. The results for TOC and UV254 do not represent daily samples, but were weekly samples collected on Wednesday.

Table 5.1

Average raw water quality during screening tests

	Week 1	Week 2
Alkalinity (mg/L as $CaCO_3$)	67	66
Ammonia (mg/L as N)	0.02	0.02
Bromide (ppb)	149	171
Hardness (mg/L as $CaCO_3$)	78	78
pH	7.6	7.7
Temperature (°C)	26	25
Turbidity (NTU)	14.3	13.2
TOC (mg/L)*	2.56	2.48
UV254 (cm^{-1})*	0.124	0.126

*Weekly sample

BROMATE FORMATION

An important benefit to CCWD associated with using ClO_2 in combination with ozone is the reduction of bromate formation in the preozonation contactor. The results of the screening tests showed that adding 0.5 mg/L or more of ClO_2 before preozonation resulted in less bromate formation than occurred when no ClO_2 was added. The results also showed that if ClO_2 is substituted for a portion of the ozone, thereby reducing the ozone dose from 1.5 mg/L to 1.0 or 0.5 mg/L, bromate formation decreased. These results suggest that preoxidation with ClO_2 is a viable method of reducing bromate formation in the preozonation contactor. Bromate formation was monitored during each of the tests by collecting samples in the pre-rapid mix, which is located downstream of the preozonation contactor. The ClO_2 dose, ozone dose, and bromate concentration for each screening test are listed in Table 5.2. The average bromate concentrations observed when test conditions were identical during the two-week test period are shown on Figure 5.1. Samples below the detection limit were treated as zero in the average calculator.

As shown on Figure 5.1, preoxidation with ClO_2 at dosages of 0.5 mg/L or higher resulted in less bromate formation than with no ClO_2 preoxidation at the same ozone dose. This trend is evident in the results achieved with ozone doses of 1.0 and 1.5 mg/L. All bromate concentrations for the 0.5 mg/L ozone dose were less than the detection limit; therefore no conclusion can be drawn from these data. The data also show that a ClO_2 dose of 0.25 mg/L actually resulted in more bromate formation than at the same ozone dose and with no ClO_2 addition. A possible reason for this anomaly is the ClO_2 reduced the ozone demand of the raw water, thus making more ozone available to react with the bromide in the raw water to form bromate. The observed increase in bromate formation following preoxidation was also observed by Zegers *et al.* 2001, who reported that an ozone dose of 1.65 mg/L resulted in bromate formation of 14 ppb, but when 0.5 mg/L chlorine was added prior to the same 1.65 mg/L ozone dose, the bromate formation increased to 21 ppb. However, increasing ClO_2 dosages (0.5 mg/L and higher) should have also decreased the ozone demand and allowed more ozone to react with bromide, but all of the tests done with 0.5 mg/L or higher dosages of ClO_2 reduced the bromate formation. Also, after ClO_2 addition, chlorite is present in the water and acts as ozone demand as the chlorite reacts with the ozone to form chlorate. So if ozone demand was increased, the bromate formation would be expected to be reduced. Reduction of bromate formation with ClO_2 at a constant ozone dose was tested in more detail during Phase 2 and the bromate mitigation testing. See these results for more discussion on the impact of ClO_2 on bromate formation.

Table 5.2
Bromate formation after preozonation during screening tests

		Week 1		Week 2	
Run	Ozone dose (mg/L)	ClO_2 dose (mg/L)	Bromate formation (ppb)	ClO_2 dose (mg/L)	Bromate formation (ppb)
1	1.0	-	1.6	-	1.3
2	1.5	-	5.7	-	7.2
3	0.5	-	<1	-	<1
4	1.0	0.25	2.8	0.5	<1
5	1.5	0.25	14	0.5	4
6	0.5	0.25	<1	0.5	<1
7	1.0	0.50	1.3	0.75	<1
8	1.5	0.50	3.2	0.75	<1
9	0.5	0.50	<1	0.75	<1
10	1.0	0.75	<1	1.0	<1
11	1.5	0.75	1.2	1.0	5.8
12	0.5	0.75	<1	1.0	<1

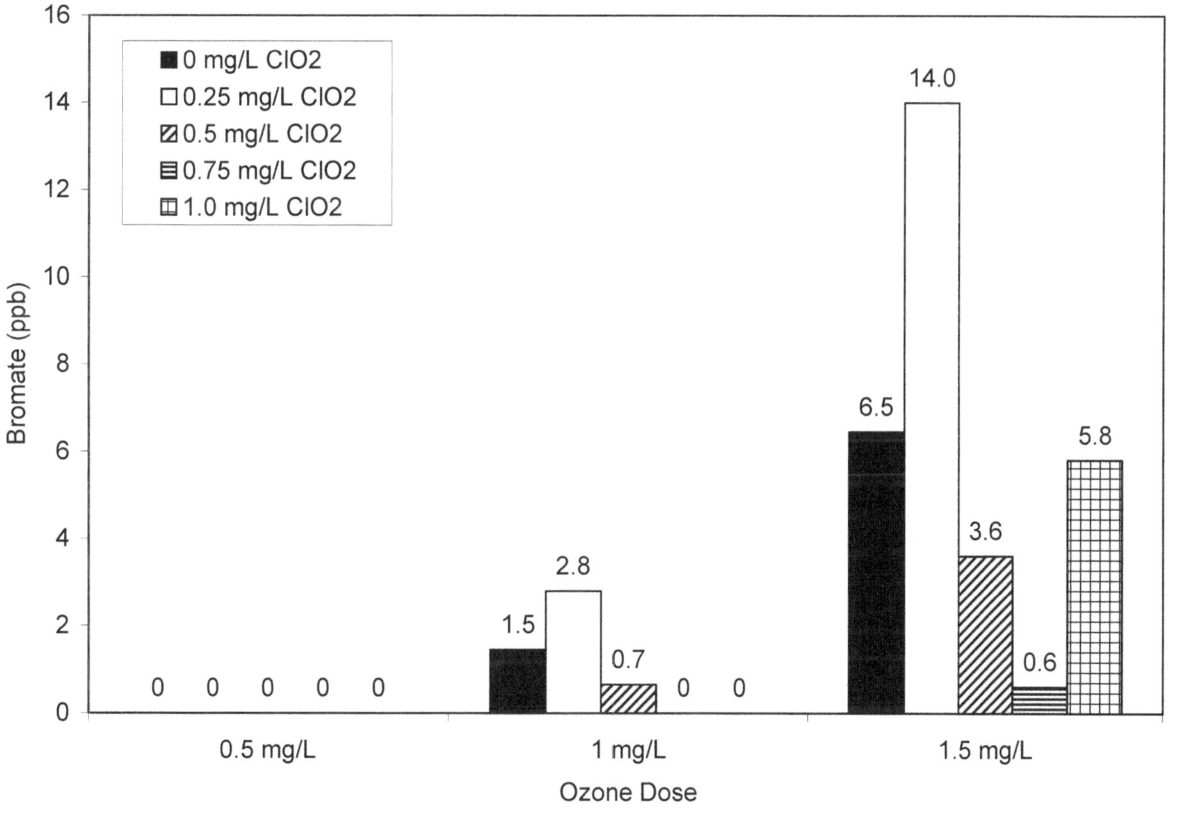

Figure 5.1 Average bromate formation during screening tests

CHLORITE/CHLORATE FORMATION

The oxidation of chlorite to chlorate by ozone was monitored during the screening tests. The results showed that an ozone dose of 1.5 mg/L oxidized virtually all of the chlorite present after ClO_2 addition. Lower ozone dosages oxidized a majority of the chlorite, but measurable amounts of chlorite remained after ozone had been added. Most of the chlorite was converted to chlorate, but when the chlorite decrease and chlorate increase were compared on a chloride ion basis, the chlorite reduction could not be completely accounted for by increases in chlorate. The differences undoubtedly could have been accounted for by increased chloride concentrations, but that was not measured.

During each test, samples were collected immediately before ozone addition at the end of Cell 1 and after ozone addition in the pre-rapid mix to monitor the oxidation of chlorite to chlorate. A summary of the chlorite concentration after ClO_2 addition, but before ozone addition, is shown on Figure 5.2. The chlorite concentration at the pre-rapid mix (after ozonation) is shown on Figure 5.3. The chlorate concentrations after ClO_2 and after ozonation are shown on Figure 5.4 and Figure 5.5, respectively. Results from weeks 1 and 2 at identical dosage conditions were averaged and shown as one data point for each of the figures.

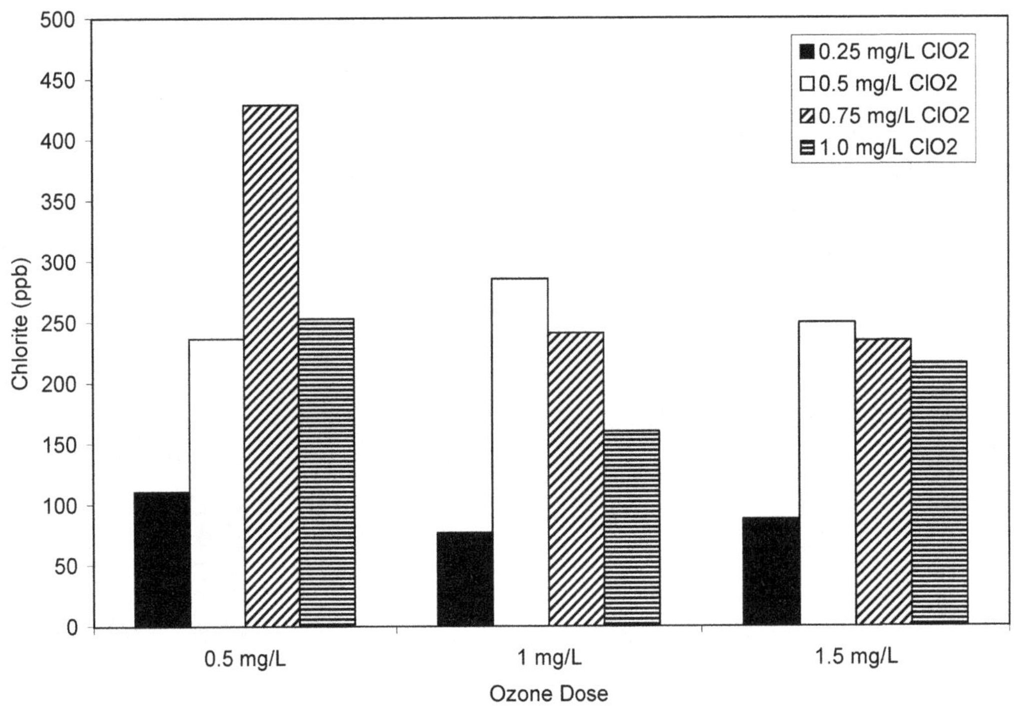

Figure 5.2 Chlorite concentrations at the end of Cell 1 after ClO_2 addition but before ozone addition in Cell 2

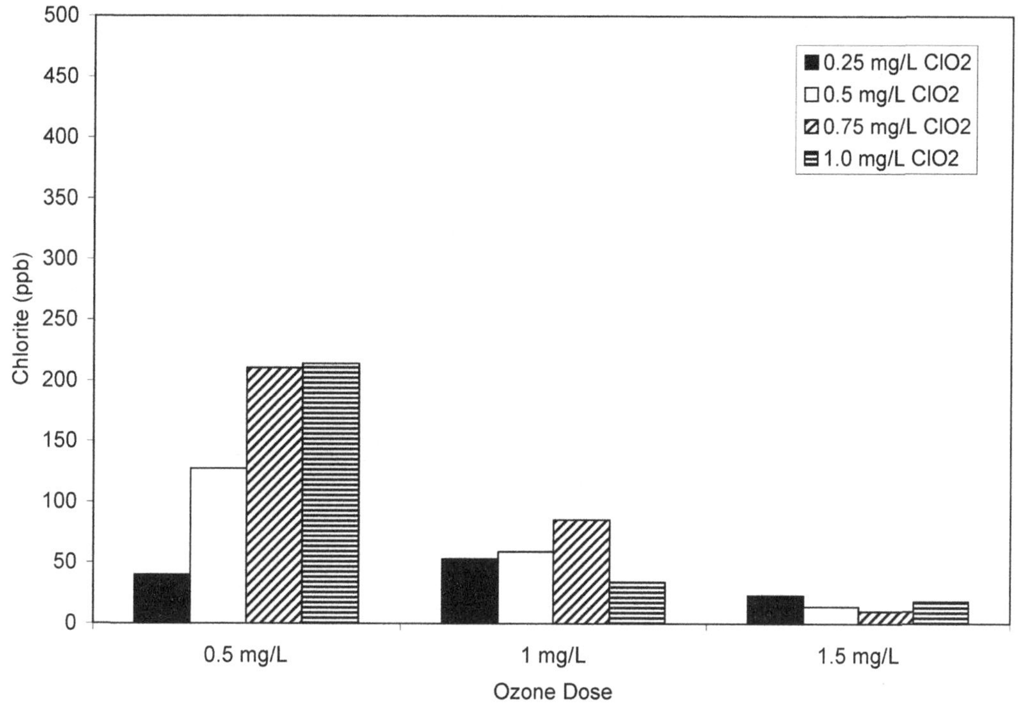

Figure 5.3 Chlorite concentrations in the pre-rapid mix after ozone addition

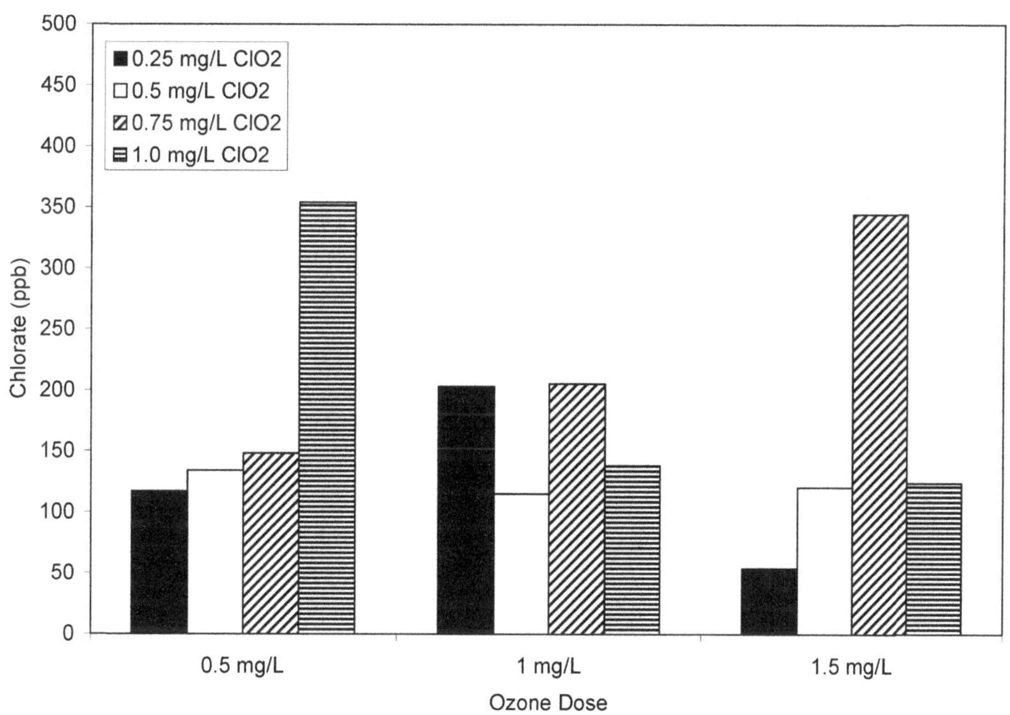

Figure 5.4 Chlorate concentrations at the end of Cell 1 after ClO_2 addition but before ozone addition in Cell 2

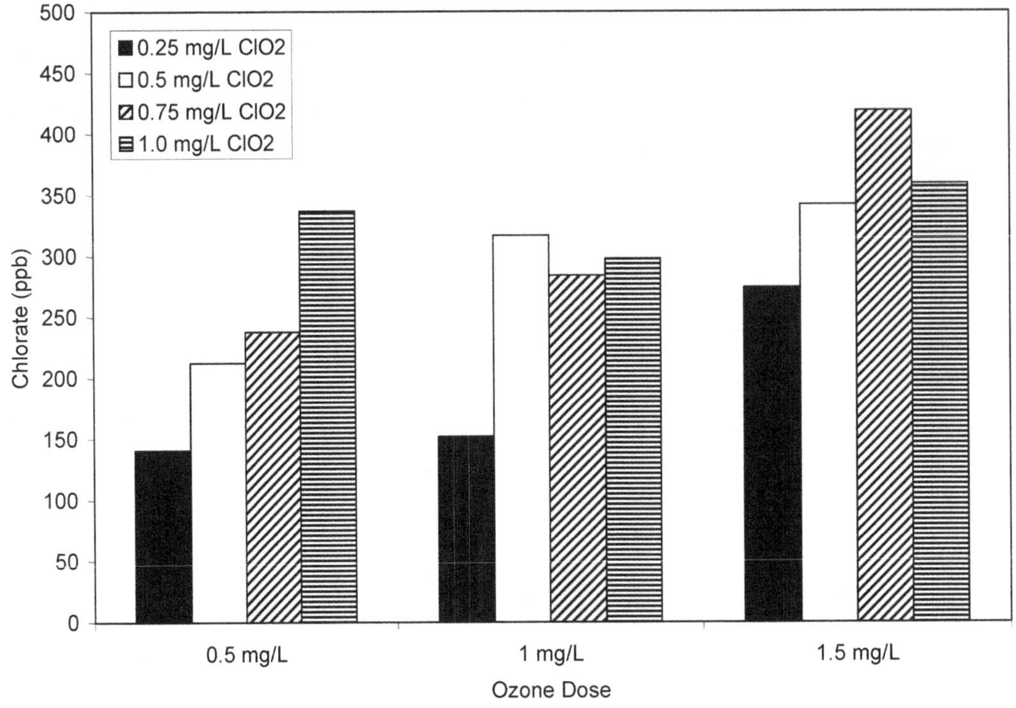

Figure 5.5 Chlorate concentrations in the pre-rapid mix after ozone addition

As shown by the data, a majority of the chlorite present after ClO_2 addition was converted to chlorate at an ozone dose of 1.5 mg/L. At lower ozone doses, chlorite was removed, but to a lesser extent. Increased removal of chlorite at higher ozone doses translated into higher chlorate concentrations after ozone. Two trends were evident: (1) at a constant ClO_2 dose, chlorate concentration increased with increasing ozone doses, and (2) at a constant ozone dose, chlorate concentration increased with increasing ClO_2 doses.

Chlorite concentrations as a percentage of the applied ClO_2 dose in many instances were lower than literature values which typically range from 30 to 70 percent of the reacted ClO_2 dose (Gates 1998). During this study, the chlorite concentration after ClO_2 addition, but before ozone addition (Figure 5.2), averaged 37 percent of the ClO_2 dose and ranged between 16 and 57 percent. No correlation was observed between the ClO_2 dose and the chlorite concentration. One explanation for the lack of correlation is that the samples collected after ClO_2 addition, but before ozone addition, were taken from the end of ozone cell 1. Since ozone is added at the beginning of cell 2, the contents of the end portion of cell 1 may have been influenced by the ozone added in cell 2. If the samples were influenced by the ozone, the chlorite concentration would be lower than expected and the chlorate concentration higher than expected, as was the case.

CHLORINE DIOXIDE AND OZONE DEMAND

Chlorine dioxide and ozone residuals were measured during each test to determine the oxidant demand of the water. The results showed very high demand for both oxidants, with little to no residuals observed. For all tests, the ClO_2 residual at the end of cell 1 was zero or very near zero. The ozone residual was zero or very near zero at the end of cell 3. The absence of residuals for both oxidants did not allow calculation of decay rates and CTs or determination of the impact of preoxidation with ClO_2 on the ozone demand.

CONCLUSIONS

Conclusions for the screening tests were as follows:

- Preoxidation with ClO_2 prior to ozonation shows promise as a bromate mitigation method at the Randall-Bold WTP. Bromate formation could be reduced by substituting ClO_2 for ozone (lowering the ozone dose).
- Ozonation following ClO_2 addition resulted in a majority of the chlorite formed being oxidized to chlorate.
- The assumption that preoxidation with ClO_2 would reduce the ozone demand could not be proved because the residuals of both oxidants were near zero during all tests.

CHAPTER 6
PHASE II – SHORT TERM TESTS

Twelve short-term tests were completed from July 27 through August 27, 2001 to refine the conclusions derived from the screening tests. The tests evaluated combining ClO_2 doses of 0.5 and 1.0 mg/L and ozone doses of 1.0 and 1.5 mg/L. In addition to evaluating the combination of ClO_2 and ozone, the short-term tests also evaluated the use of ClO_2 only as a preoxidant to determine if the plant was able to maintain filtration enhancement without preozonation. One test was conducted with no preoxidation, to serve as a baseline for all the tests. Detailed information on the testing plan can be found in Chapter 4. The following chapter describes the bromate formation, chlorite/chlorate formation, SDS testing, ClO_2 and ozone demands, and presents the conclusions derived from the data.

BROMATE FORMATION

The bromate concentration was monitored after both preozonation and postozonation to determine the impact of the preoxidation with ClO_2. The data from the testing indicated that at a constant ozone dose, preoxidation with ClO_2 resulted in less bromate formation in both preozonation and postozonation contactors than occurred when no ClO_2 was added. The data also suggest that the test with ClO_2 preoxidation only resulted in a lower finished water bromate concentration than the test with no preoxidation. When ozone alone was used for preoxidation, the preozonation contactor contributed the majority of the bromate formation, since it had a higher dose than the postozonation contactor, and the ammonia used for the formation of chloramines was added ahead of the postozonation contactor.

Samples for bromate measurement for each condition were collected at the end of each 48-hour test. The samples were taken from the pre-rapid mix (after preozonation) and from the effluent of the postozonation contactor. During the testing, the preoxidant doses were maintained constant during the test by the plant control system, and the postozonation dose was varied as needed to ensure compliance with primary disinfection requirements. A summary of the ClO_2, preozonation, and postozonation doses; the postozonation CT; and the bromate concentration after preozonation and postozonation addition are listed in Table 6.1.

As shown in Table 6.1, less bromate was formed in both the preozonation and postozonation contactors when ClO_2 was added than the tests without ClO_2 addition. The data from tests 2 and 3 show that with ClO_2 preoxidation (no preozonation), the bromate formation in the postozonation contactor was lower than when no preoxidation was used. For tests with ozone only, a majority of the bromate formation occurred in the preozonation contactor. This was expected, since the postozonation doses are typically lower and ammonia (for chloramine formation) was added ahead of the postozonation contactor.

When the tests were compared on the basis of equal preozonation doses, preoxidation with ClO_2 showed favorable results for bromate formation. The average bromate concentrations for all tests with 1.0 mg/L of preozonation and varying ClO_2 doses are shown on Figure 6.1.

Table 6.1
Bromate formation during short-term tests

Test Number	ClO_2 Dose (mg/L)	Pre-O_3 Dose (mg/L)	Post-O_3 Dose* (mg/L)	Post-O_3 CT* (mg*min/L)	Bromate after Pre-O_3 (ppb)	Bromate after Post-O_3 (ppb)
1	-	1.0	0.53	4.95	6.0	8.2
2	-	-	0.72	3.19	-	2.9
3	1.0	-	0.83	3.38	<2	<2
4	-	1.0	0.54	3.35	9.9	7.8
5	1.0	1.0	0.47	3.93	1.2	2.4
6	0.5	1.5	0.46	3.76	6.4	14.4
7	-	1.5	0.45	3.40	12	11.7
8	0.5	1.0	0.47	3.36	<2	3.7
9	1.0	1.0	0.50	3.65	<2	<2
10	-	1.0	0.48	3.59	4.4	7.5
11	0.5	1.0	0.48	3.23	<2	<2
12	-	1.0	0.54	3.37	5.7	7.5

*Measured at end of testing during sampling period.

Although using ClO_2 to lower the ozone dose reduced the bromate formation during the screening tests, the data on Figure 6.1 show that preoxidation with ClO_2 actually reduced the bromate formation when the ozone dose was not changed. The addition of 1.0 mg/L of preozonation resulted in an average bromate formation of 6.5 ppb after preozonation and 7.8 ppb after postozonation. However, adding 0.5 or 1.0 mg/L of ClO_2 before preozonation reduced the average bromate formation to less than 0.6 ppb after preozonation and to less than 1.9 ppb after postozonation. This discovery was not expected during the initial planning of the project. The ability of ClO_2 to reduce the bromate formation at a constant ozone dose will be further confirmed and discussed in more detail in Chapter 8.

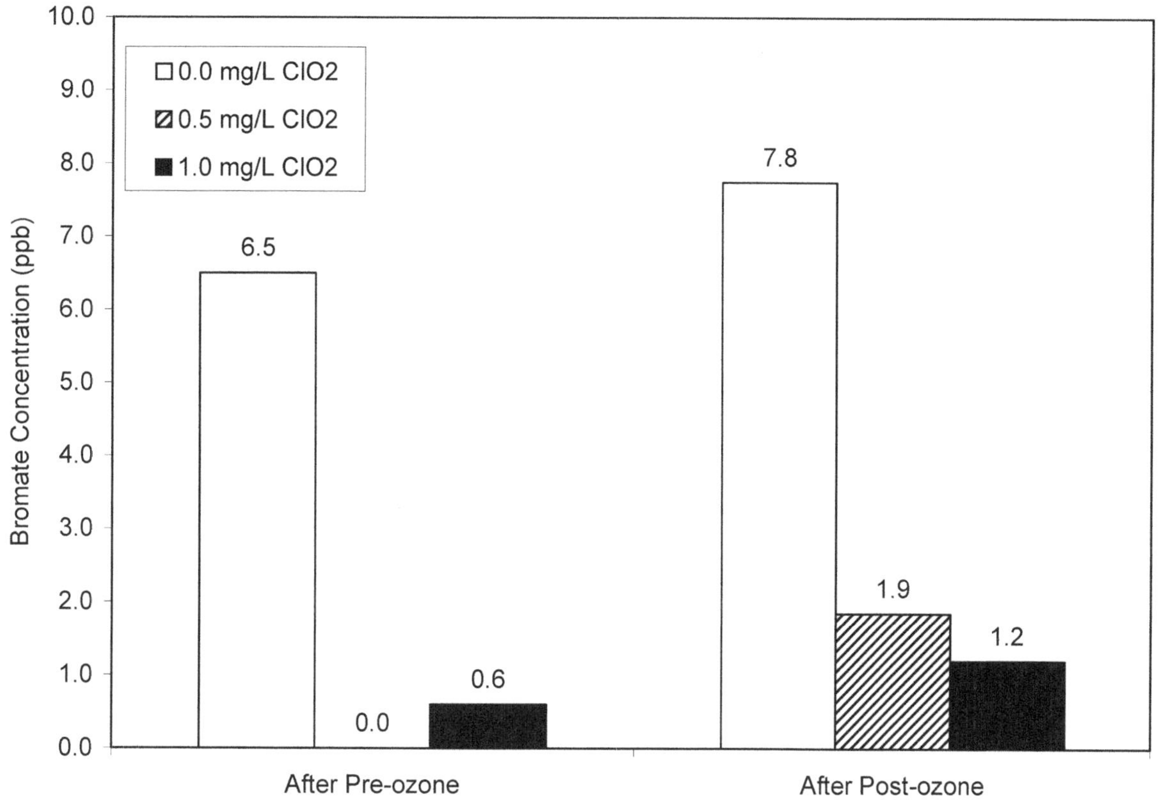

Figure 6.1 Average bromate formation during short-term tests with a 1.0 mg/L preozonation dose

CHLORITE/CHLORATE FORMATION

The chlorite and chlorate concentrations were monitored during testing to determine the extent of chlorite formation after ClO_2 addition, and of chlorate formation as a result of chlorite reacting with ozone. The results showed that 18 to 69 percent of the ClO_2 that was added contributed to the formation of chlorite. Virtually all of the chlorite was removed in the preozonation step, where the chlorite was oxidized to chlorate, therefore, the chlorite concentration in the plant effluent was below detection in all the tests. The results from test run 3 showed that the GAC filters were able to remove chlorite without increasing the chlorate concentration.

Measurements for chlorate and chlorite concentrations were made once per run, near the end of each test run. Samples were collected after ClO_2 addition, after preozonation, from the GAC filter effluent, and from the plant effluent. A summary of the chlorite and chlorate concentrations for each of the test runs is listed in Table 6.2.

Table 6.2
Chlorite and chlorate concentrations during short-term tests

			Chlorite (ppb)				Chlorate (ppb)			
Test Number	ClO_2 Dose (mg/L)	O_3 Dose (mg/L)	After ClO_2	After Preozonation	GAC filter effluent	Plant effluent	After ClO_2	After Preozonation	GAC filter effluent	Plant effluent
1	-	1.0	-	ND	-	ND	-	ND	-	ND
2	-	-	-	-	-	ND	-	-	-	ND
3	1.0	-	286	279	44	ND	47	54	51	179
4	-	1.0	-	ND	-	ND	-	ND	-	ND
5	1.0	1.0	180	ND	ND	ND	34	250	250	270
6	0.5	1.5	210	ND	ND	ND	54	290	200	250
7	-	1.5	-	-	-	-	-	-	-	-
8	0.5	1.0	346	ND	ND	ND	60	464	446	430
9	1.0	1.0	623	29	ND	ND	98	892	860	860
10	-	1.0	-	ND	-	ND	-	ND	-	ND
11	0.5	1.0	325	ND	ND	ND	63	446	408	502
12	-	1.0	-	ND	-	ND	-	ND	-	ND

The chlorite data show that 18 to 69 percent of the ClO_2 added was converted to chlorite. As noted previously in this report, typical chlorite concentrations after ClO_2 addition range from 30 to 70 percent of the reacted ClO_2 dose (Gates 1998). After ozonation, the majority of the chlorite was oxidized to chlorate by ozone. For the tests with the combination of ClO_2 and ozone, this led to finished water chlorite concentrations below the detection limit and chlorate concentrations varying from 250 to 860 ppb. Test run 9 showed a high degree of chlorite formation which led to a finished water chlorate concentration that exceeded the California action level of 800 ppb. The high chlorate formation at Randall-Bold would limit the ClO_2 dosage so as not to exceed the chlorate AL.

During test run 3, when ClO_2 was added without preozonation, the GAC filters removed the majority of chlorite, and the chlorate concentration did not increase. Chlorite removal by the GAC filters resulted in the lowest effluent chlorate concentration for any of the short-term tests with ClO_2 addition. Hoehn et al. [N.d.] found that GAC removal of chlorite diminished over time. The long-term testing will provide better data on the chlorite capacity of the GAC filters.

SDS TESTING

To determine the effect of preoxidation on DBP formation, SDS tests were conducted on both raw and finished water to measure the THM and HAA formation potential. Samples were collected during five of twelve short-term tests. A summary of the test conditions and the SDSTHM and SDSHAA concentrations for the raw and finished water are presented in Table 6.3. Detailed information on the conditions used for the SDS tests can be found in Chapter 3. The results during the short-term tests were inconclusive because of limited sampling and limited replicates. Data from the long-term tests (Chapter 7) provided a better basis for comparison than the limited tests that were conducted during the short term.

Table 6.3
DBP formation during short-term testing

Test Number	ClO_2 Dose (mg/L)	O_3 Dose (mg/L)	Influent SDS TTHM (ppb)	Filter Effluent SDS TTHM (ppb)	Influent SDS HAA_5 (ppb)	Filter Effluent SDS HAA_5 (ppb)
5	1.0	1.0	136	61	34	14
7	0.0	1.5	173	56	47	12
8	0.5	1.0	173	60	47	14
11	0.5	1.0	163	73	37	29
12	-	1.0	177	80	41	26

CHLORINE DIOXIDE AND OZONE DEMAND

Chlorine dioxide and ozone residuals were measured during each test to determine the oxidant demand of the water. The results showed very high demand for both oxidants, with little to no residuals observed. For all tests, the ClO_2 residual at the end of cell 1 was zero or very near zero. The ozone residual was zero or very near zero at the end of cell 3. The absence of residuals of both oxidants did not allow calculation of decay rates and CTs or determination of the impact of preoxidation with ClO_2 on the ozone demand.

CONCLUSIONS

Conclusions from the Phase 2 testing were:

- Preoxidation with ClO_2 prior to ozonation can either substantially reduce or eliminate entirely the formation of bromate.
- At the Randall-Bold WTP, all of the chlorite that was formed after ClO_2 addition was either oxidized to chlorate or removed by the GAC.
- The GAC filter can reduce the chlorite concentration, while chlorate is neither removed nor produced.

CHAPTER 7
PHASE III – LONG-TERM TESTS

To provide full-scale operating data for evaluating ClO_2 preoxidation, three-week long-term tests were conducted from August 27, 2001 through January 7, 2002. The results showed that ClO_2 preoxidation lowered the bromate formation during subsequent ozonation and reduced the total ozone use while maintaining filtered water particulate removal. The chlorite that was formed after ClO_2 addition was either converted to chlorate by reaction with ozone or removed in the GAC filters. During testing, three preoxidation conditions were evaluated: 1.5 mg/L of ClO_2 only, 1.0 mg/L of ClO_2 with 0.5 mg/L of ozone, and 1.0 mg/L of ozone only. Each of the three conditions was tested for a period of about three weeks and then the sequence was repeated, providing over 4 months of full-scale data. The test conditions and dates of operation are listed in Table 7.1; more detailed information on the test procedure and sampling is presented in Chapter 4.

Table 7.1

Long-term testing conditions

Phase	Starting Date	Ending Date	ClO_2 Dose (mg/L)	Ozone Dose (mg/L)
3A	8/27/01	9/16/01	1.5	-
3B	9/17/01	10/8/01	1.0	0.5
3C	10/13/01	11/4/01	-	1.0
3D	11/7/01	11/26/01	1.5	-
3E	11/29/01	12/18/01	1.0	0.5
3F	12/21/01	1/7/02	-	1.0

RAW WATER QUALITY

The raw water quality was monitored daily for several constituents during the long-term testing. During the first three months, the raw water quality was consistent, but it changed markedly during the final month of testing. The turbidity, ammonia concentration, TOC concentration, and UV254 adsorption increased steadily over the final month of testing while the pH decreased over the same period. The raw water quality changes during December were attributed to changes that were made to the source water in the canal during that month.

Raw water was sampled daily in accordance with the sampling protocol; analyses included temperature, flow rate, turbidity, pH, hardness, alkalinity, ammonia, TOC, and UV254. Data summarizing the raw water quality are presented on Figure 7.1 through Figure 7.6.

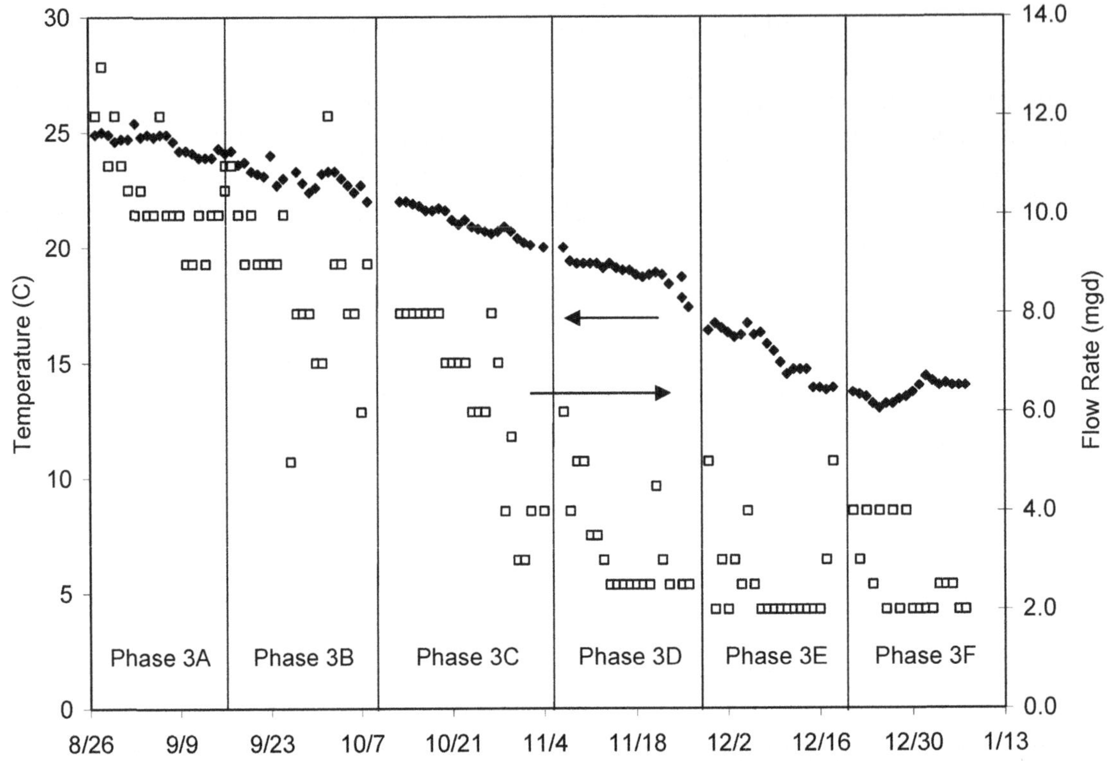

Figure 7.1 Raw water temperature and flow rate during long-term testing

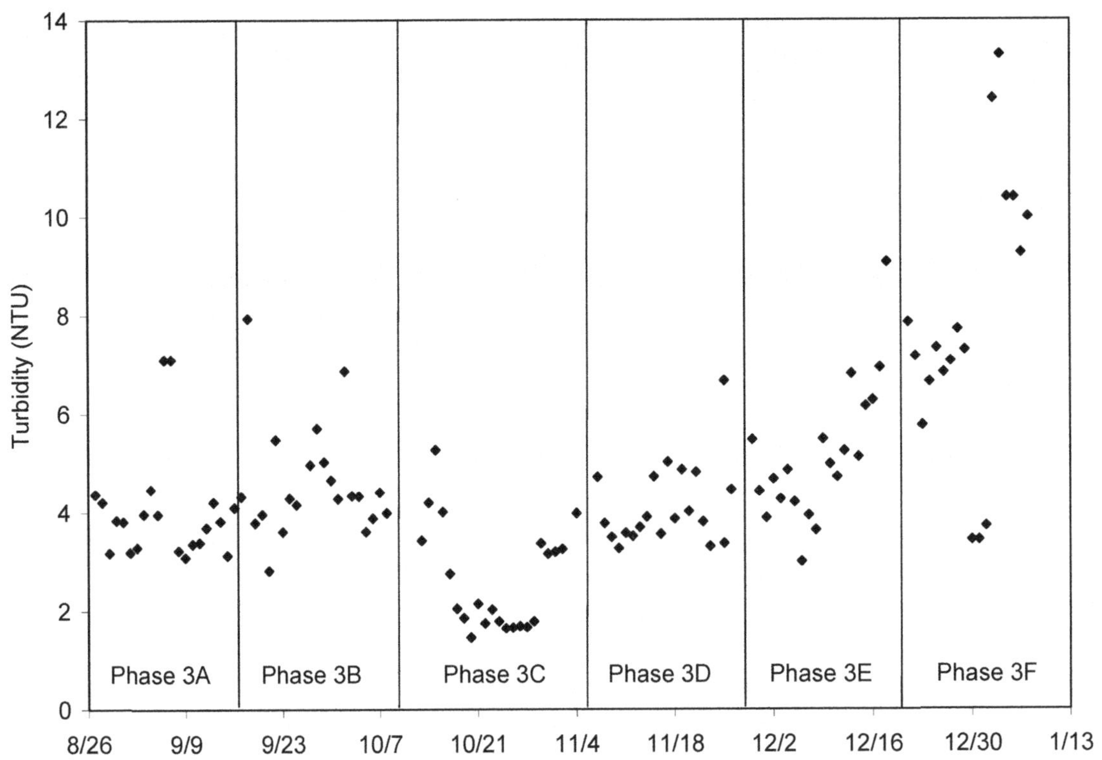

Figure 7.2 Raw water turbidity during long-term testing

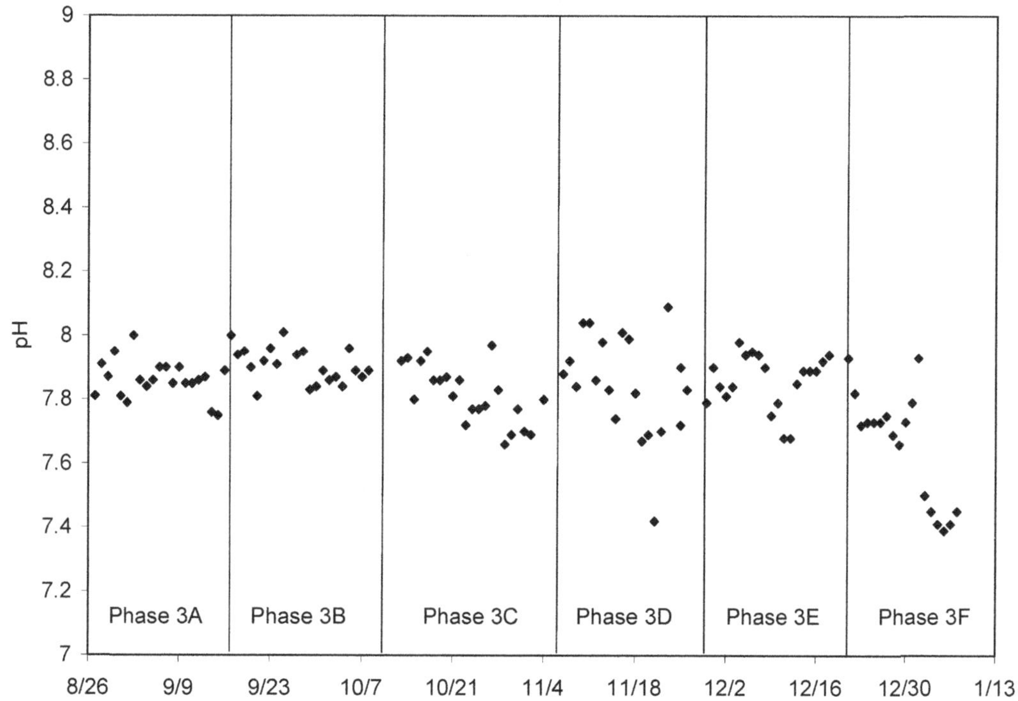

Figure 7.3 Raw water pH during long-term testing

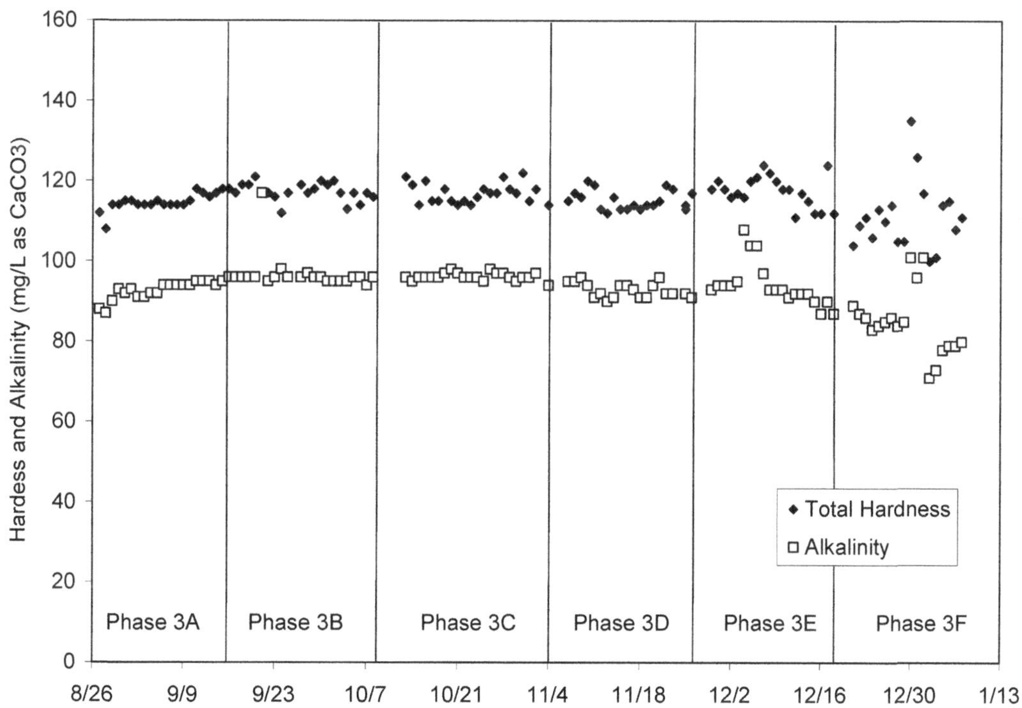

Figure 7.4 Raw water total hardness and alkalinity during long-term testing

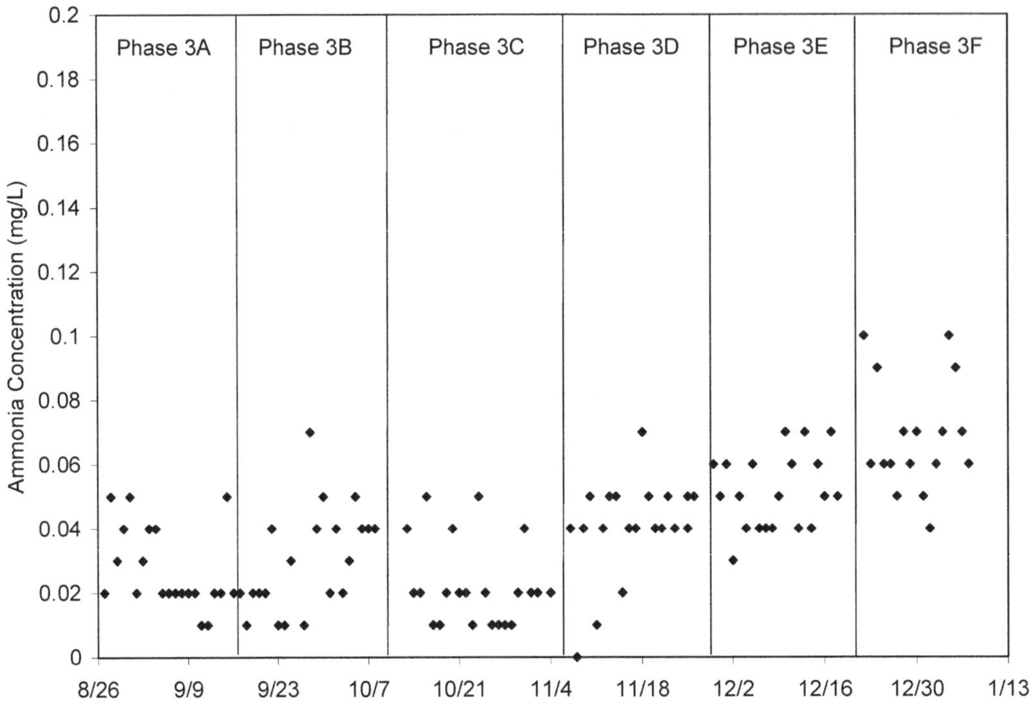

Figure 7.5 Raw water ammonia concentration during long-term testing

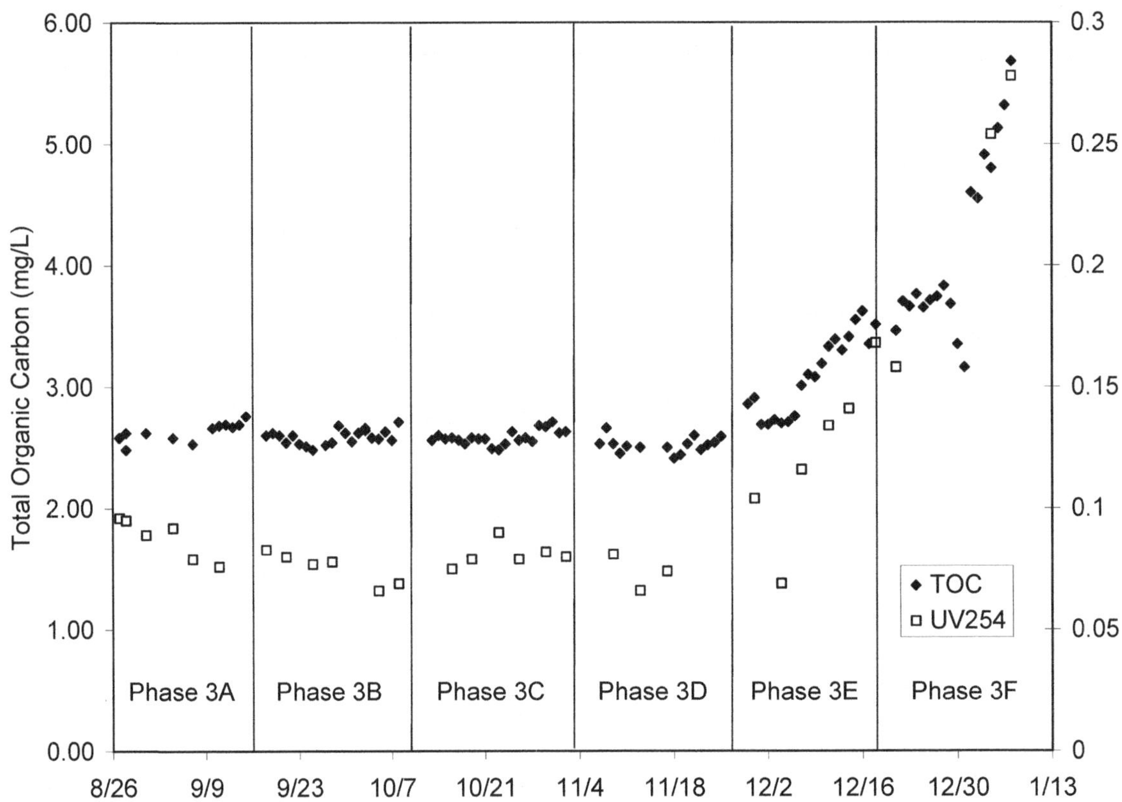

Figure 7.6 Raw water TOC and UV254 during long-term testing

As shown on the preceding figures, the raw water quality was relatively consistent during the first three months of testing, but changed substantially over the last month. Beginning in December, turbidity steadily increased from around 4 NTU to around 12 NTU; the pH decreased from about 7.8 units to about 7.4 units, while the total hardness and alkalinity decreased slightly; the ammonia concentration increased slightly. The TOC concentration and the UV254 adsorbance showed a substantial increase: the TOC concentration increased from about 2.6 mg/L to almost 6 mg/L by the end of the month and the UV254 absorbance increased from about 0.07 cm^{-1} to 0.28 cm^{-1}.

The observed changes in raw water quality during December were the result of nine changes to the source water in the canal that were made during that month. Source water blending for the canal is a complex process that takes into account many factors such as pumping costs, fish migration, and chloride concentration goals. The changes during December gradually reduced the contribution from the Los Vaqueros Reservoir from 85 percent prior to December to zero percent of the blended source water at the end of December. The Los Vaqueros contribution was replaced with water derived directly from the two direct delta sources, Rock Slough and Old River, both of which generally are more turbid and contain higher concentrations of organic matter.

BROMATE FORMATION

During the long-term full-scale testing, bromate formation was monitored to evaluate the possible benefits of ClO_2 preoxidation. The results showed that ClO_2 preoxidation reduced the finished water bromate concentration when compared to the control tests done without ClO_2. Tests with ClO_2 preoxidation resulted in bromate concentrations less than the detection limit of 1 ppb in all the samples after preozonation, and in average finished water bromate concentrations less than 2 ppb. On average, the two test conditions with ozone alone, however, resulted in bromate concentrations of 6.2 and 3.9 ppb in the finished water.

The bromate concentration was measured twice a week in the pre-rapid mix and in the filter effluent, postozonation effluent, and plant effluent. On the days of bromate sampling, the raw water bromide concentration was also measured for comparison. The average bromate concentrations, average raw water bromide concentration, ClO_2 dose, preozonation dose, and average postozonation dose for each testing condition are listed in Table 7.2. Bromate concentrations below the detection limit of 1 ppb were treated as zero in the calculation of averages. The bromate concentration in the raw water was measured once per week, but none was detected.

Table 7.2
Impact of preoxidation on postozonation dose and bromate formation

Phase	Average raw water bromide (ppb)	ClO_2 dose (mg/L)	Pre-O_3 dose (mg/L)	Average post-O_3 dose (mg/L)	Average post-O_3 CT (mg·min/L)	Average* bromate after pre-O_3 (ppb)	Average* bromate after filtration (ppb)	Average* bromate after post-O_3 (ppb)	Average* bromate finished water (ppb)
3A	265	1.5	-	1.22	4.26	0	0	0.2	1.5
3B	257	1.0	0.5	0.78	3.99	0	0	0.8	1.9
3C	305	-	1.0	0.54	3.94	4.8	3.9	5.6	6.2
3D	343	1.5	-	1.53	4.97	0	0	0	0.6
3E	316	1.0	0.5	1.16	5.16	0	0	0.5	2.0
3F	255	-	1.0	1.13	5.14	0.3	1.1	3.2	3.9

* Concentrations below the detection limit were treated as zero in the average calculation

During the long-term testing, the average raw water bromide concentration varied from 255 to 343 ppb. This concentration of bromide is very high compared to the concentration in most waters; the average bromide concentration for all plants in the ICR was 90 ppb and the highest concentration for the plants using ozone was 300 ppb (Rice *et al.* 2000). Although the average raw water bromide concentrations varied some from phase to phase, this should not have effected the bromate formation substantially.

As shown in Table 7.2, the tests with ClO_2 preoxidation resulted in lower bromate concentrations in the preozonation, postozonation, and plant effluent when compared to the tests without ClO_2. Phases 3A and 3D, which added 1.5 mg/L of ClO_2 alone, had the lowest bromate formation with average finished water concentrations of 1.5 and 0.6 ppb, respectively. The tests done with a combination of ClO_2 and ozone, Phases 3B and 3E, had slightly higher finished water bromate concentrations of 1.9 and 2.0 ppb. The tests with ozone alone, Phases 3C and 3F, had the highest finished water bromate concentrations of 6.2 and 3.9 ppb. Adding ClO_2 reduced the total ozone dose (preozonation dose plus postozonation dose) which is likely one factor in reducing bromate formation. Subsequent special tests (Chapter 8) have also shown that ClO_2 addition before ozonation limits bromate by a mechanism that is not currently understood.

The decreased bromate formation during Phase 3F compared to Phase 3C is likely the result of raw water quality changes over the last month of testing. During Phase 3F, the raw water turbidity increased, the pH decreased, the ammonia concentration increased, and TOC concentration increased from about 3 mg/L to 6 mg/L. The increase in turbidity more than likely increased the ozone demand, which should decrease the bromate formation. Decreasing the pH or increasing the ammonia concentration are both methods for bromate formation control, so these changes likely decreased the bromate formation. Increasing TOC concentrations with constant ozone doses (or decreasing the O_3/DOC ratio) has also been shown to decrease bromate formation (Amy and Siddiqui 1999).

CHLORITE/CHLORATE FORMATION

The chlorite and chlorate concentrations were monitored at several points in the plant to determine the amount of chlorite that formed after ClO_2 addition and the amount that was oxidized to chlorate by ozone. The results showed that there were three chlorite removal processes in the treatment train and that effluent chlorite concentrations were typically below the detection limit of 10 ppb. Both the preozonation and postozonation converted a majority of the chlorite to chlorate. The GAC filters typically removed about half of the chlorite present in the filter influent; however, unlike ozone, the GAC removed the chlorite without causing the chlorate concentration to increase.

The chlorite and chlorate concentrations were measured twice a week in the raw water, post-ClO_2, pre-rapid mix, filter effluent, postozonation, and plant effluent. Average concentrations were calculated for each phase and concentrations below the detection limit were treated in the calculation as zero. The average chlorite and chlorate concentrations for Phases 3A and 3D, 1.5 mg/L of ClO_2 alone, are shown on Figure 7.7 and Figure 7.8. The average chlorite and chlorate concentrations for Phases 3B and 3E, 1.0 mg/L of ClO_2 and 0.5 mg/L of ozone, are shown on Figure 7.8 and Figure 7.10. Since the chlorite and chlorate concentrations in the raw water were always below the detection limits, during Phases 3C and 3F they were not measured.

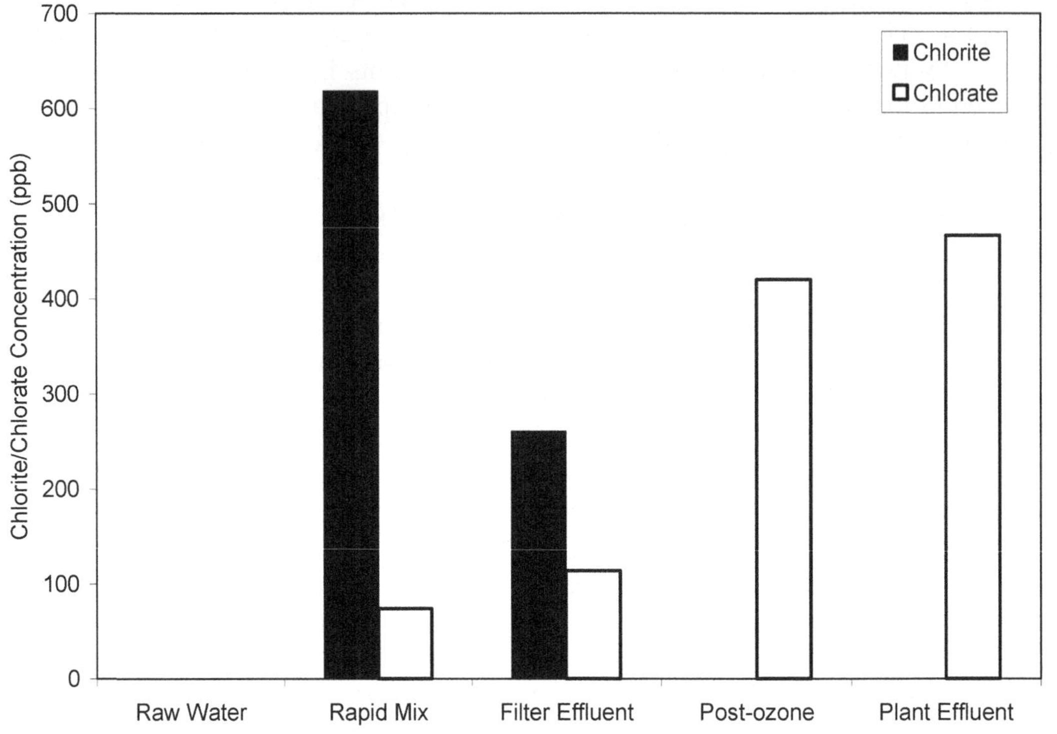

Figure 7.7 Average chlorite and chlorate concentrations during Phase 3A – 1.5 mg/L ClO_2 alone

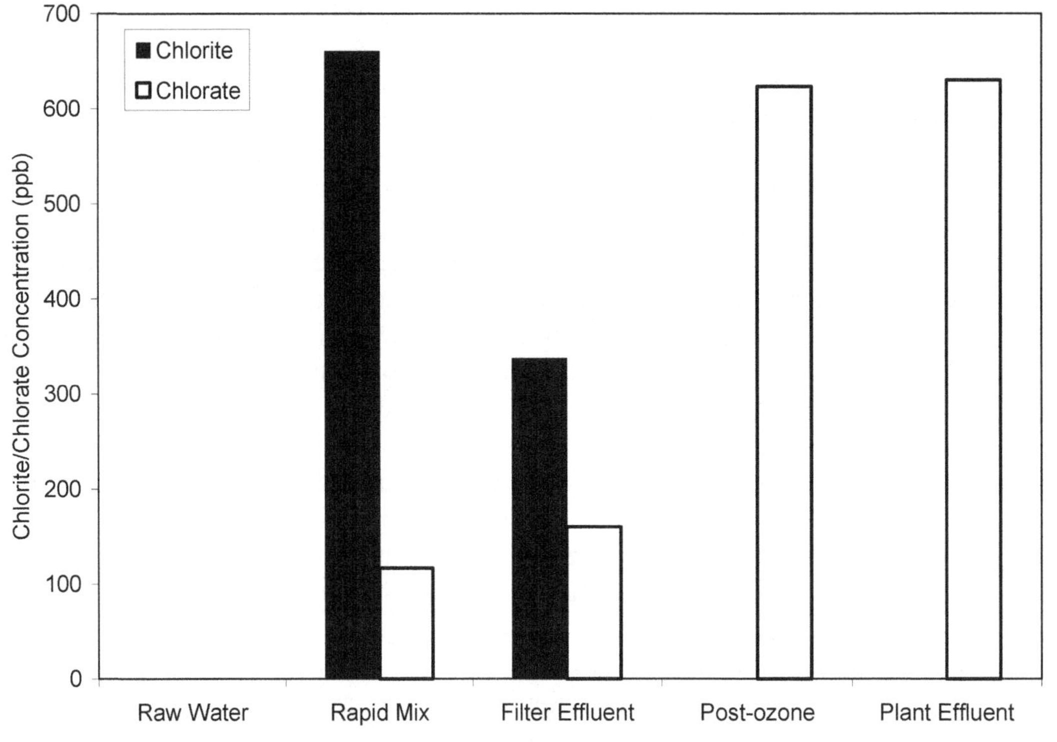

Figure 7.8 Average chlorite and chlorate concentrations during Phase 3D – 1.5 mg/L ClO_2 alone

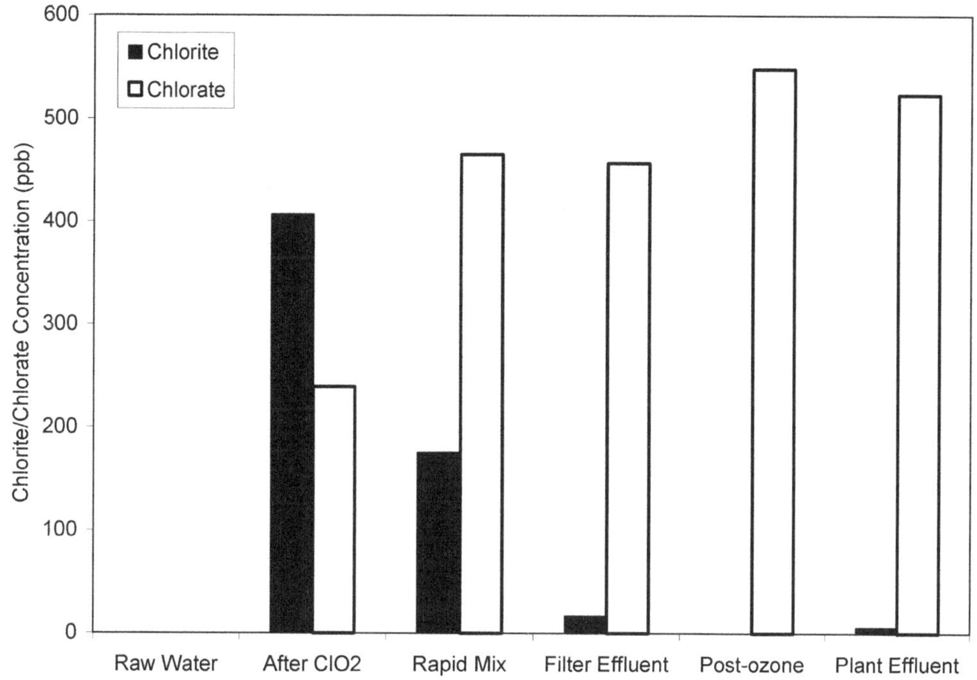

Figure 7.9 Average chlorite and chlorate concentrations during Phase 3B – 1.0 mg/L ClO_2 and 0.5 mg/L ozone

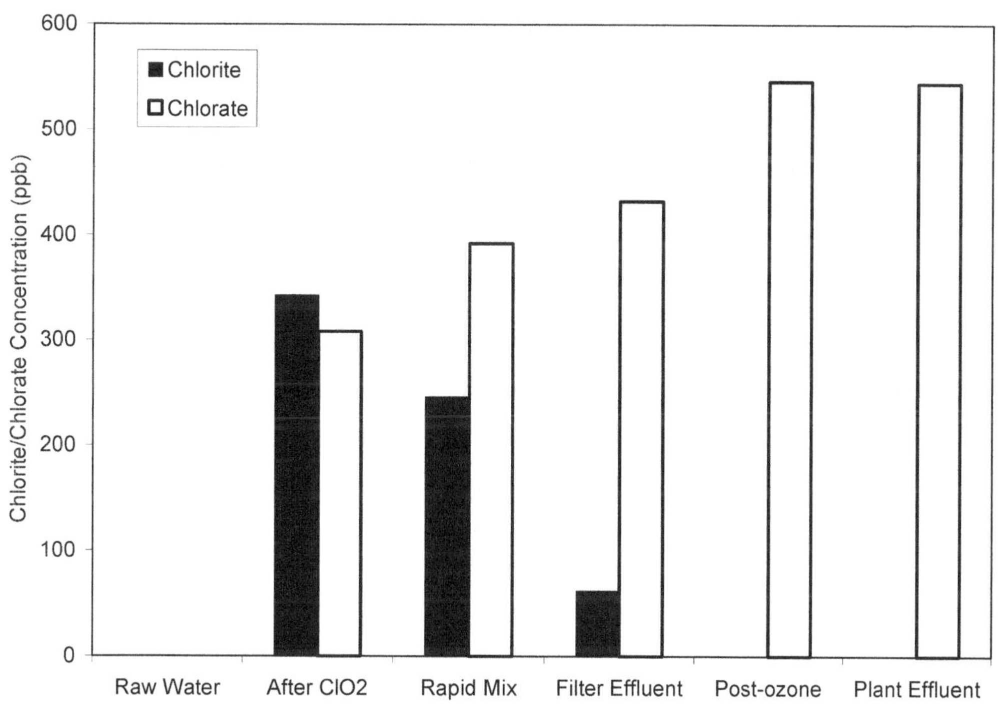

Figure 7.10 Average chlorite and chlorate concentrations during Phase 3E – 1.0 mg/L ClO_2 and 0.5 mg/L ozone

During Phases 3A and 3D, when 1.5 mg/L of ClO_2 alone was used for preoxidation, the chlorite concentration in the pre-rapid mix averaged 640 ppb, which is 43 percent of the applied ClO_2 dose. About half of the chlorite formed was then removed in the GAC filters, with only a small increase in the chlorate concentration across the filters. Other work has shown that chlorite is mostly converted to chloride by the GAC and chlorate was only formed when chlorine was also present (Minear and Amy 1996). Postozonation oxidized the remaining chlorite to chlorate, so that the chlorite concentration in the plant effluent was always below the detection limit of 10 ppb. Since some of the chlorite was oxidized to chlorate, the chlorate concentration increased through the plant with a plant effluent average of about 550 ppb.

The chlorite and chlorate profiles through the treatment plant were different during Phases 3B and 3E, when 1.0 mg/L ClO_2 was applied in combination with 0.5 mg/L of ozone. The average chlorite concentration after ClO_2 addition, but before preozonation, was 375 ppb, which was 37 percent of the applied ClO_2 dose. The preozonation then oxidized about half of the chlorite to chlorate and the GAC filters removed most of the rest. The small amount of the chlorite that remained in the filter effluent was oxidized to chlorate by the postozonation, so that the plant effluent chlorite concentration was typically below the detection limit and the average chlorate concentration was 534 ppb.

The chlorite concentration in the filter effluent during each phase appeared to directly influence the required postozonation dose (Table 7.2). Chlorite exerts ozone demand and the more chlorite in the filter effluent, the more ozone that will be needed to maintain residuals for disinfection credit. Based on the stoichiometry, 1 mg/L of chlorite would consume 0.71 mg/L of ozone (Minear and Amy 1996). The postozonation doses for Phases 3A through 3C can be compared because they all have a similar CT value of about 4 mg*min/L and the doses from Phases 3D through 3F can be compared because they all have a similar CT value of about 5 mg*min/L. During all four phases of testing that included ClO_2, the required postozonation dose for disinfection was higher than in the two phases without ClO_2. The tests with 1.5 mg/L ClO_2 alone resulted in the highest chlorite concentrations in the filter effluent, and therefore with the highest postozonation doses. The tests with a combination of ClO_2 and ozone resulted in lower chlorite concentrations in the filter effluent, and the required postozonation dose was lower than in the tests with ClO_2 alone, but higher than in the tests without ClO_2 addition (which resulted in no chlorite in the filter effluent). Besides the presence of chlorite, the reduced preozonation dosage could have also influenced the required postozonation doses.

CHLORINE DIOXIDE AND OZONE DEMAND

The ClO_2 and ozone residuals were monitored daily during long-term operation to determine the ClO_2 demand and the impact that ClO_2 preoxidation has on ozone demand. The results showed that the initial ClO_2 residual concentration (measured 62 feet down stream of the addition point) was highly variable: during testing with a 1.5 mg/L dose of ClO_2, the initial residual varied from zero to 1.2 mg/L, and with a 1.0 mg/L dose of ClO_2, the initial residual varied from zero to 0.5 mg/L. The variability of the initial residual is more than likely caused by changes in the raw water quality caused by changes in the blend of water in the canal. The impact of ClO_2 on the ozone demand could not be determined, because virtually all of the ozone residual concentrations at the end of Cell 2 were zero or very near zero.

The ClO_2 residual was measured daily during long-term operation from XI-106 (initial residual), post-ClO_2 (end of Cell 1), the end of Cell 3, and the pre-rapid mix. The measurements were made using the LGB method; however, beginning with Phase 3B, the initial ClO_2 residual was also measured using the one-step amperometric method for comparison. The initial ClO_2 residual values during Phases 3A, 3B, 3D, and 3E are summarized on Figure 7.11 through Figure 7.14.

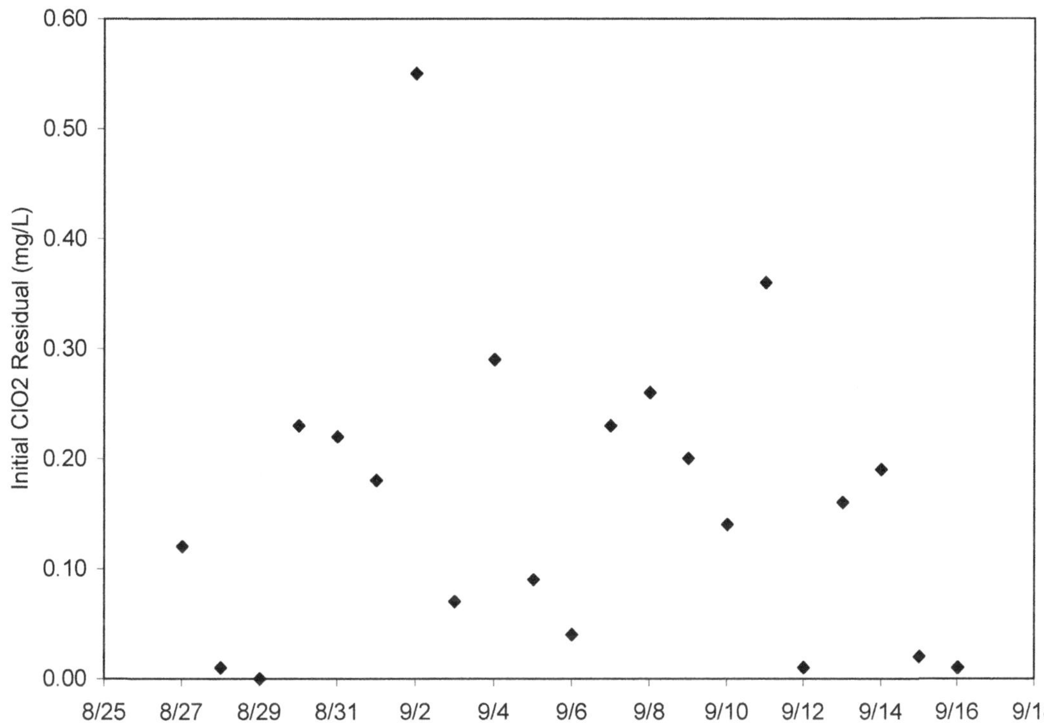

Figure 7.11 Initial ClO_2 residual during Phase 3A, 1.5 mg/L ClO_2 dose

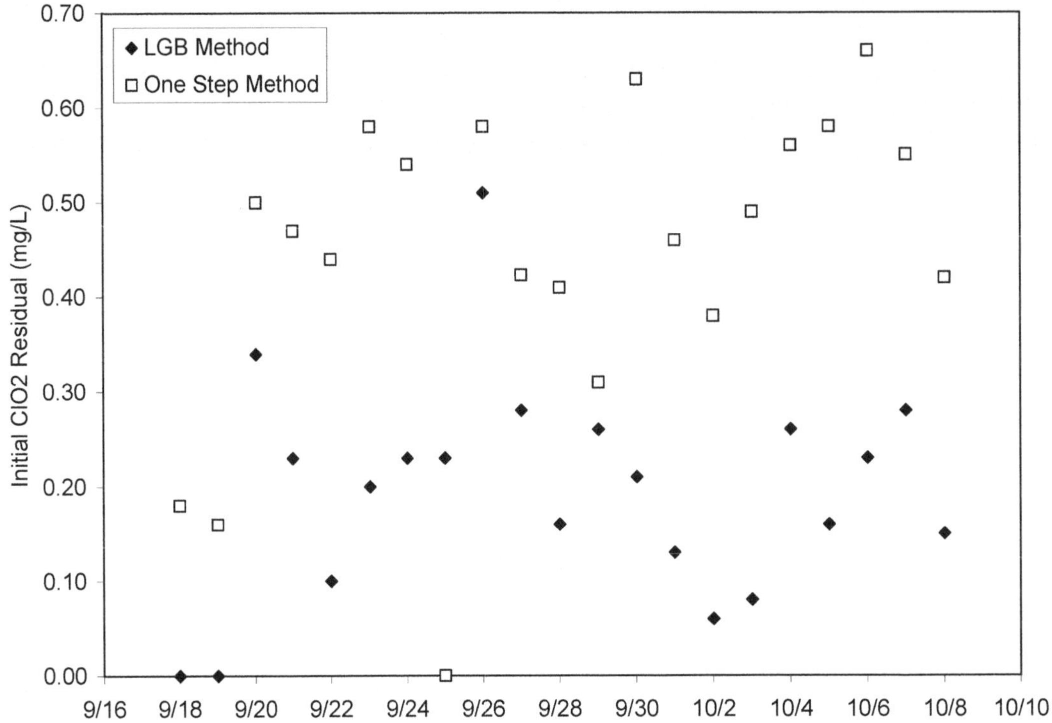

Figure 7.12 Initial ClO_2 residual during Phase 3B, 1.0 mg/L ClO_2 dose

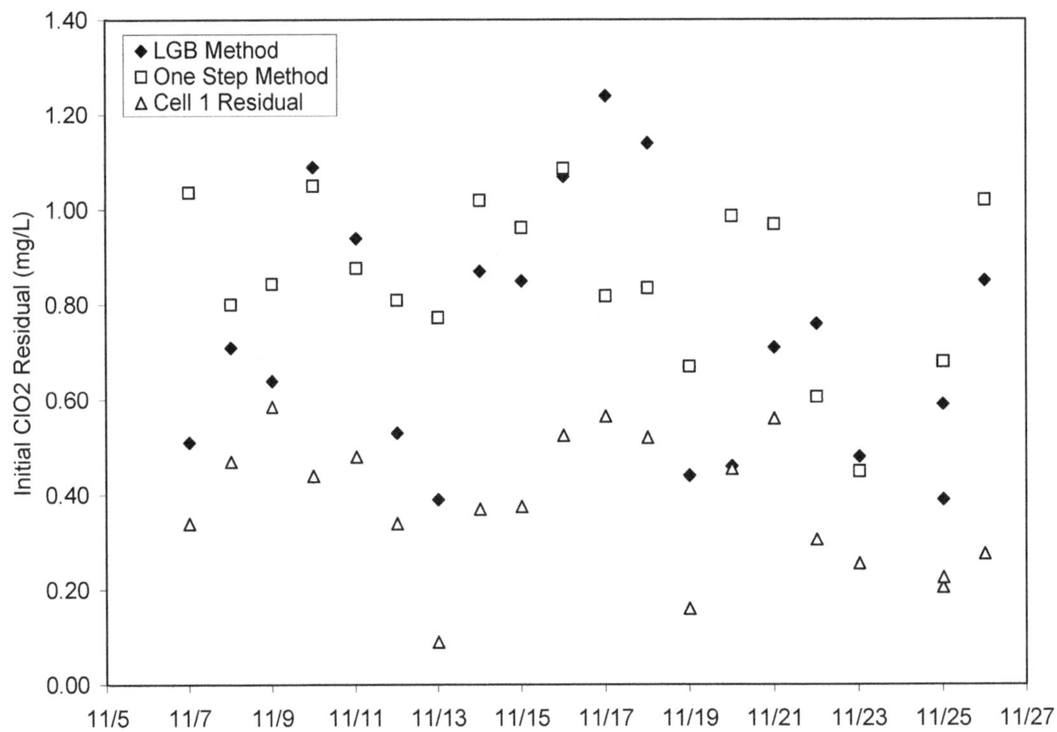

Figure 7.13 Initial ClO_2 residual during Phase 3D, 1.5 mg/L ClO_2 dose

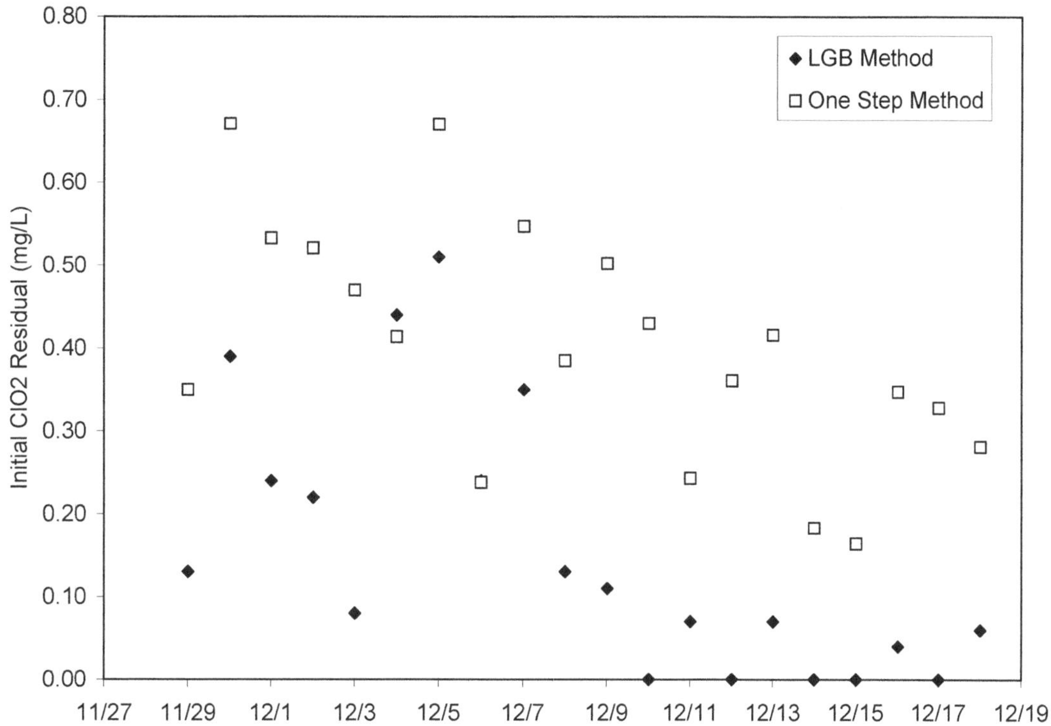

Figure 7.14 Initial ClO_2 residual during Phase 3E, 1.0 mg/L ClO_2 dose

As shown on the previous figures, the initial ClO_2 residual varied widely from day to day. During Phase 3A, when 1.5 mg/L of ClO_2 was added, the initial residual varied from zero to 0.55 mg/L and during Phase 3D, it varied from 0.4 mg/L to 1.2 mg/L. During Phases 3B and 3E, when 1.0 mg/L of ClO_2 was added before 0.5 mg/L of ozone, the initial residual varied from zero to 0.5 mg/L. Daily changes in raw water quality and oxidant demand is the most likely explanation for the highly variable ClO_2 residual. The daily variability cannot be confirmed either from past or current ozone data because the ozone demand in the raw water is so high that only minimal ozone residuals are observed in the cell after ozone addition.

Incomplete mixing in the 48-inch raw water pipeline following ClO_2 addition was considered as another possible reason for the highly variable ClO_2 residuals but was deemed less likely than changes in the raw water quality. The initial sampling point was about 62 feet downstream from the feed point, which in a 48-inch diameter pipeline is a distance of about 15 pipe diameters. A general rule of thumb for mixing is a distance of at least 10 pipe diameters. However, the same day to day variability is evident when the ClO_2 residual at the end of Cell 1, which would be even more than 10 pipe diameters down stream of the dosing point, is plotted with the initial residual (Figure 7.13). So, although some of the variability in the initial residual could be attributed to incomplete mixing, the majority of the variability can be attributed to other causes, such as raw water quality changes.

The ClO_2 residual was also measured at several points in the treatment process to develop residual profile curves. The average ClO_2 residuals at XI-106, at the end of Cell 1, at the end of Cell 3, and at the pre-rapid mix for each phase are shown on Figure 7.15.

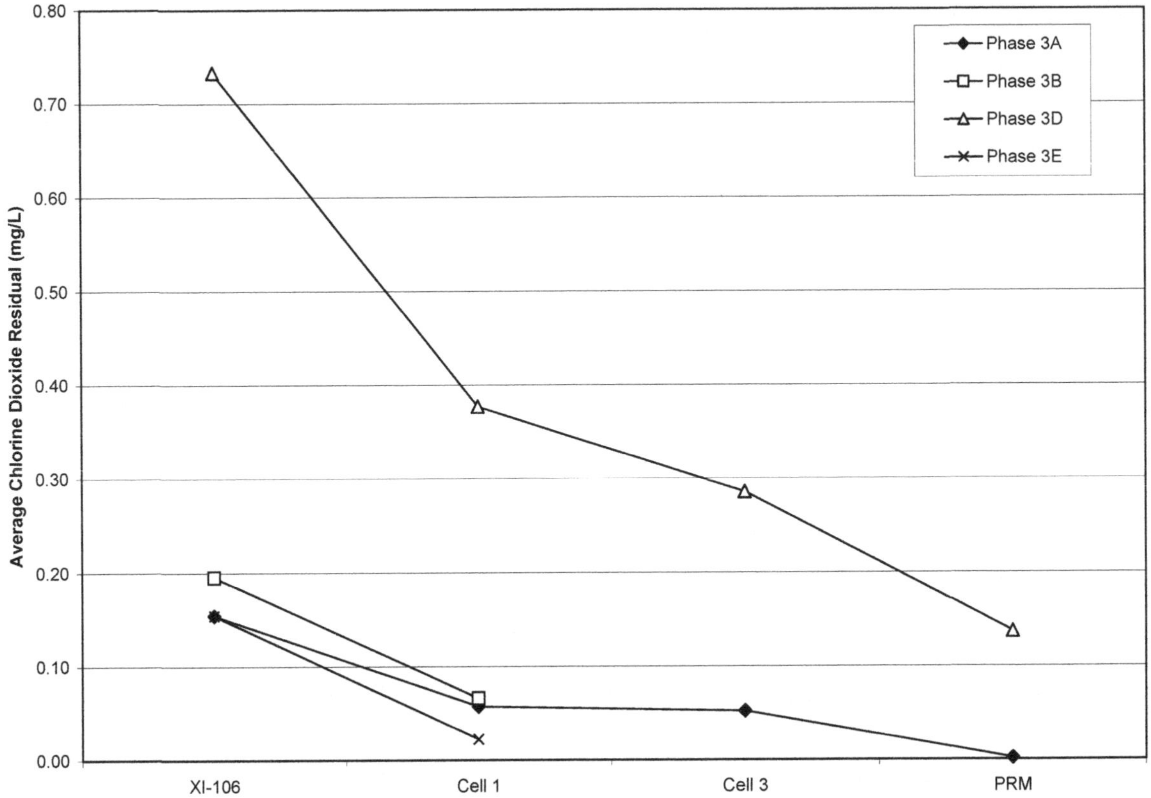

Figure 7.15 Average ClO_2 residual during long-term testing (LGB method).

Although the ClO_2 doses used in Phases 3A and 3D were the same, the average residual profiles through the plant were substantially different. The 1.5 mg/L ClO_2 dose during Phase 3D resulted in an average initial residual of 0.73 mg/L and an average residual at the pre-rapid mix, after the preozonation contactors, of 0.14 mg/L. During Phase 3A, the initial residuals were much lower, averaging only 0.15 mg/L. One possible explanation for the difference in ClO_2 residuals is raw water temperatures were about 25°C during Phase 3A and 20°C during Phase 3D, therefore, the ClO_2 may have decayed more rapidly in the warm water making it difficult to get measurable residuals. For the testing with a combination of ClO_2 and ozone, the ClO_2 dose was reduced to 1.0 mg/L and the average initial residual dropped to 0.15 mg/L and 0.20 mg/L for Phases 3B and 3E, respectively. By the end of Cell 1, just before ozone addition, the ClO_2 residual then decayed to about 0.05 mg/L. Thus, the majority of the ClO_2 had disappeared before ozone was added in Cell 2.

The impact of the ClO_2 addition on the preozonation demand could not be directly measured. During all phases of the testing with ozone, the initial ozone residual (at the end of Cell 2) was zero or virtually zero. Therefore, no direct correlation could be made on how much the ClO_2 reduced the preozonation demand of the raw water. The ClO_2 addition did influence the postozonation demand (or dose), as discussed previously.

COAGULATION AND FILTRATION IMPACT

An important factor in substituting preoxidation with ClO_2 for ozone is that it should not have an adverse impact on the performance of the coagulation and filtration process. To evaluate the potential impact of substituting ClO_2, the coagulant doses, filtered water turbidities, TOC removal, and DBP formation potential were monitored during the long-term testing. The results showed that although the preoxidation scheme was rotated among three conditions, the coagulant and polymer dosages did not need to be substantially changed and there was little difference in filter effluent turbidities. In terms of organics removal, the removal of TOC and DBP formation potential for the three preoxidation conditions were also nearly the same. Therefore, it appears that ClO_2 could be substituted for ozone without adverse impact on the coagulation and filtration process.

Coagulant Doses

The coagulant doses used were recorded daily to determine whether substituting ClO_2 for preozonation would affect coagulation and require higher coagulant doses. The data show that there was not a substantial difference in the coagulant, coagulant aid, or filter aid dosages under the different testing conditions. The changes that were observed were more likely the result of changes in source water quality than of the preoxidation scheme.

Alum was used as the primary coagulant and was fed at the flocculation basin inlet. Mixing was accomplished by throttling the inlet valve to produce a 1 foot head loss. A cationic polymer (Sweetwater 110), added at the end of the flocculation basin, was used as a coagulant aid. A non-ionic polymer (Sweetwater 2992 PWG), used as a filter aid, was added at the in-line flash mix without any flash mix water. A summary of the testing conditions and the average doses of alum, cationic polymer, and non-ionic polymer is presented in Table 7.3. The daily alum dosages during the long-term testing are shown on Figure 7.16.

Table 7.3

Average coagulant and polymer doses during long-term testing

Phase	ClO_2 dose (mg/L)	Preozonation dose (mg/L)	Average alum dose (mg/L)	Average cationic polymer dose (mg/L)	Average non-ionic polymer dose (mg/L)
3A	1.5	-	25.45	1.00	0.014
3B	1.0	0.5	25.00	1.00	0.010
3C	-	1.0	20.64	1.00	0.010
3D	1.5	-	20.00	1.00	0.010
3E	1.0	0.5	24.00	1.00	0.010
3F	-	1.0	31.39	1.00	0.010

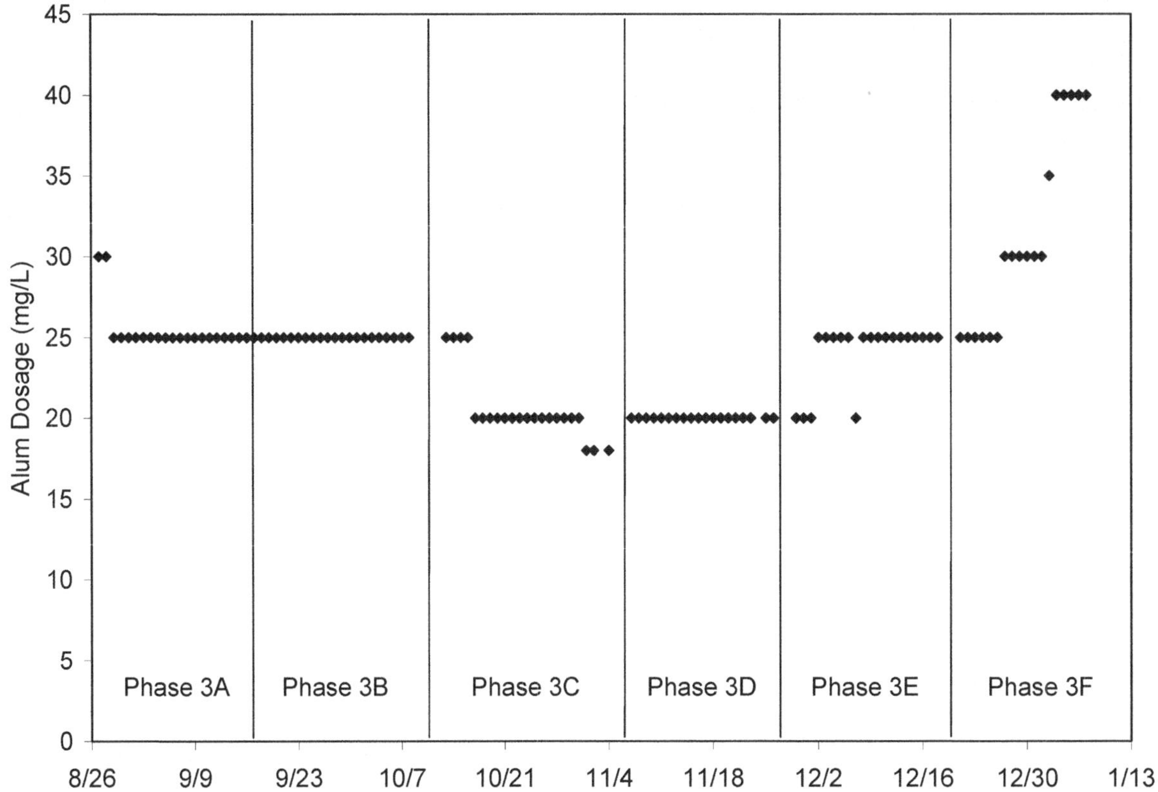

Figure 7.16 Alum dosage during long-term testing

As shown in Table 7.3, the average dosages of cationic and non-ionic polymers were almost identical during all the phases and required no adjustment when preoxidation conditions were changed. Some changes were made to the alum dosage, as shown on Figure 7.16. The increase in alum dosage during Phases 3E and 3F was the result of changes in the raw water quality rather than changes in preoxidation, as previously discussed. Note in Figure 7.16 that the alum dose remained relatively constant during Phase 3A when only ClO_2 was added and Phase 3B when both oxidants were added. While it is evident that the alum dose decreased during Phase 3C when ozone was the only preoxidant, it remained constant during the next phase when ClO_2 was substituted for ozone. Therefore, it appears that the preoxidation conditions tested did not require a substantial change in the coagulant and polymer dosages.

Filtered Water Turbidity

The turbidity at the effluent from each filter was monitored during the long-term testing to determine whether changing the preoxidation conditions would have an impact on the filter performance. The results showed that during Phases 3A through 3C, when raw water quality was essentially the same, there was no substantial difference in the effluent turbidity from all filters. Making a direct comparison of Phases 3D through 3F was difficult because there was a substantial change in raw water quality. Therefore, it appears that ClO_2 could be partially or wholly substituted for ozone without a substantial impact on the filter effluent turbidity.

To compare filter performance, effluent turbidity data were downloaded from the plant's SCADA system at 15 minute intervals and transferred to Excel for data analysis. The filter turbidity data from all testing phases are not directly comparable because it was not collected at the same time in parallel conditions with the same raw water quality. However, some comparison can be made between phases as long as the raw water quality remained relatively unchanged. Comparison of turbidity data of individual filter runs from one phase with those from another phase would be difficult, and the results would not be consistent.

Therefore, to compare preoxidation conditions, effluent turbidity data from individual filters were grouped by phase and in order from smallest to largest values to establish percentile rankings. Two percentiles (50th and 95th) were chosen for the comparison. The 50th percentile values essentially represent the median values and the 95th percentile values are important because future regulations will require the filter effluent turbidity to be less than 0.3 NTU 95 percent of the time (Federal Register 1998). The 50th percentile values for the combined, average, and individual filter effluent turbidity are listed in Table 7.4 and the 95th percentile turbidity values are listed in Table 7.5. The percentile ranking establishes the percentages of turbidity values that are less than a given turbidity. For example, during Phase 3A, 50 percent of the turbidity values from Filter 1 were less than 0.040 NTU (Table 7.4) and 95 percent were less than 0.044 NTU (Table 7.5). The data were also plotted as percentiles versus turbidity. The combined, average, and individual filter effluent values are shown on Figure 7.17 through Figure 7.24.

Examination of the filter turbidity data reveals an obvious disagreement between the individual filter effluent data and the combined filter effluent data (Table 7.4). The general ranking of the phases based on the individual filter effluent from the lowest to the highest turbidity would be as follows: Phase 3B, 3A, 3C, 3D, 3F, and 3E. For the combined filter effluent, the ranking of phases from lowest to highest turbidity would be: Phase 3C, 3A, 3F, 3E, 3D, and 3B. This analysis indicates that conditions during Phase 3B produced the lowest turbidity values based on the individual filter effluent data, but the highest values based on the combined filter effluent data. The combined filter effluent values do fall within the range of the individual filter values during Phases 3A, 3C, and 3F, but are considerably higher for the remaining three phases. The reason for this discrepancy has not been determined.

As an alternate method of characterizing the combined filter effluent, an "average" filter effluent value was calculated by averaging the on-line filters from the 15 minute turbidity data. It was not strictly an average of all six filters. For example, at a given time, if filters 1, 4, and 6 were on-line, the average filter effluent turbidity was calculated by averaging the turbidity from only those three filters. This created an average filter effluent value every 15 minutes for the entire phase of testing. The data were then sorted and percentile rankings were calculated in the same manner as was used for analyzing the other filter data.

Table 7.4

Filter effluent turbidity (NTU) - 50th percentile

Phase	Filter #1	Filter #2	Filter #3	Filter #4	Filter #5	Filter#6	Average* Filter Effluent	Combined Filter Effluent
3A	0.040	0.029	0.031	0.030	0.034	0.027	0.032	0.036
3B	0.037	0.028	0.030	0.028	0.034	0.027	0.031	0.078
3C	0.040	0.029	0.033	0.027	0.035	0.029	0.033	0.035
3D	0.047	0.032	0.036	0.035	0.041	0.030	0.035	0.069
3E	0.055	0.046	0.046	0.034	0.046	0.040	0.045	0.062
3F	0.054	0.033	0.043	0.038	0.048	0.043	0.043	0.048

*Average filter effluent represents the average of the in-service filters every 15 minutes

Table 7.5

Filter effluent turbidity (NTU) - 95th percentile

Phase	Filter #1	Filter #2	Filter #3	Filter #4	Filter #5	Filter#6	Average* Filter Effluent	Combined Filter Effluent
3A	0.044	0.032	0.037	0.036	0.041	0.034	0.038	0.044
3B	0.041	0.032	0.033	0.030	0.038	0.030	0.035	0.140
3C	0.054	0.032	0.040	0.036	0.041	0.031	0.042	0.042
3D	0.051	0.035	0.040	0.041	0.044	0.034	0.043	0.091
3E	0.063	0.050	0.055	0.045	0.048	0.046	0.059	0.091
3F	0.068	0.038	0.048	0.047	0.053	0.052	0.056	0.056

*Average filter effluent represents the average of the in-service filters every 15 minutes

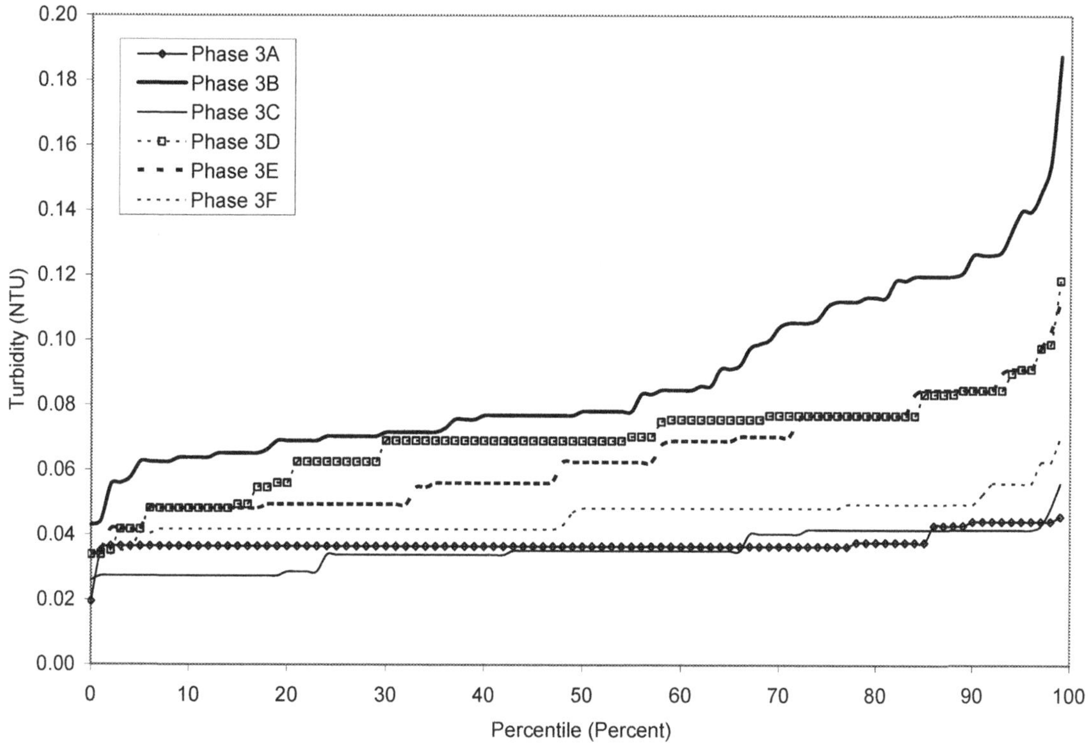

Figure 7.17 Turbidity percent less than plot – combined filter effluent

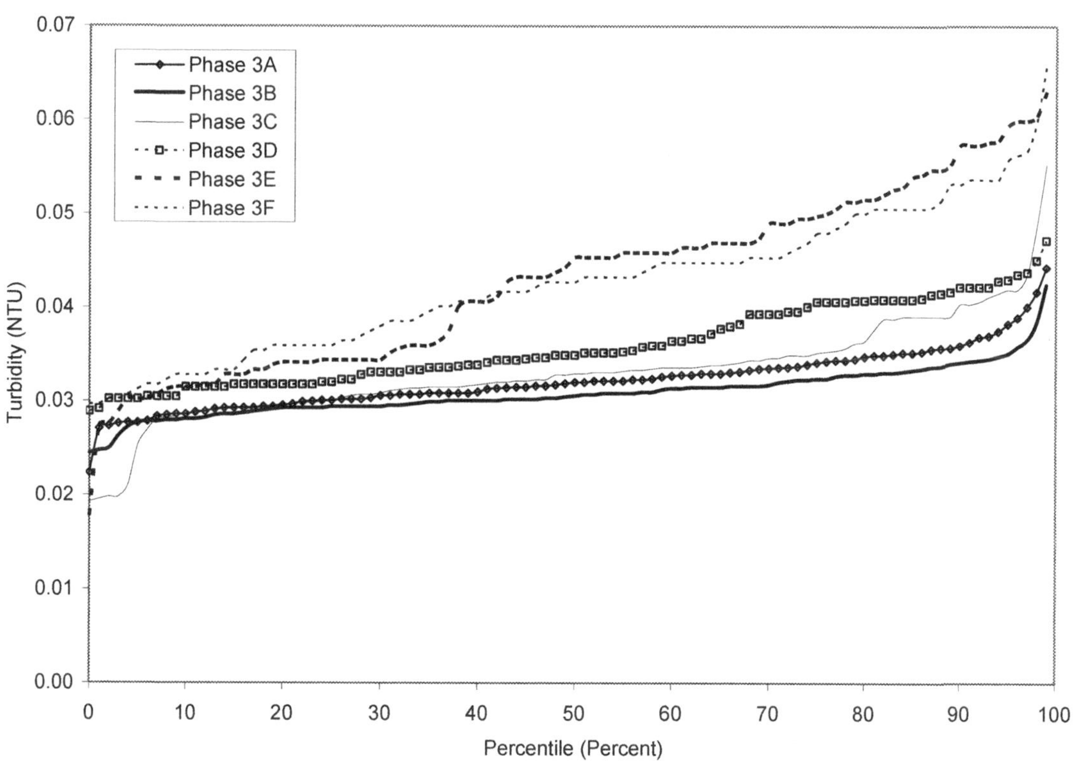

Figure 7.18 Turbidity percent less than plot – average filter effluent

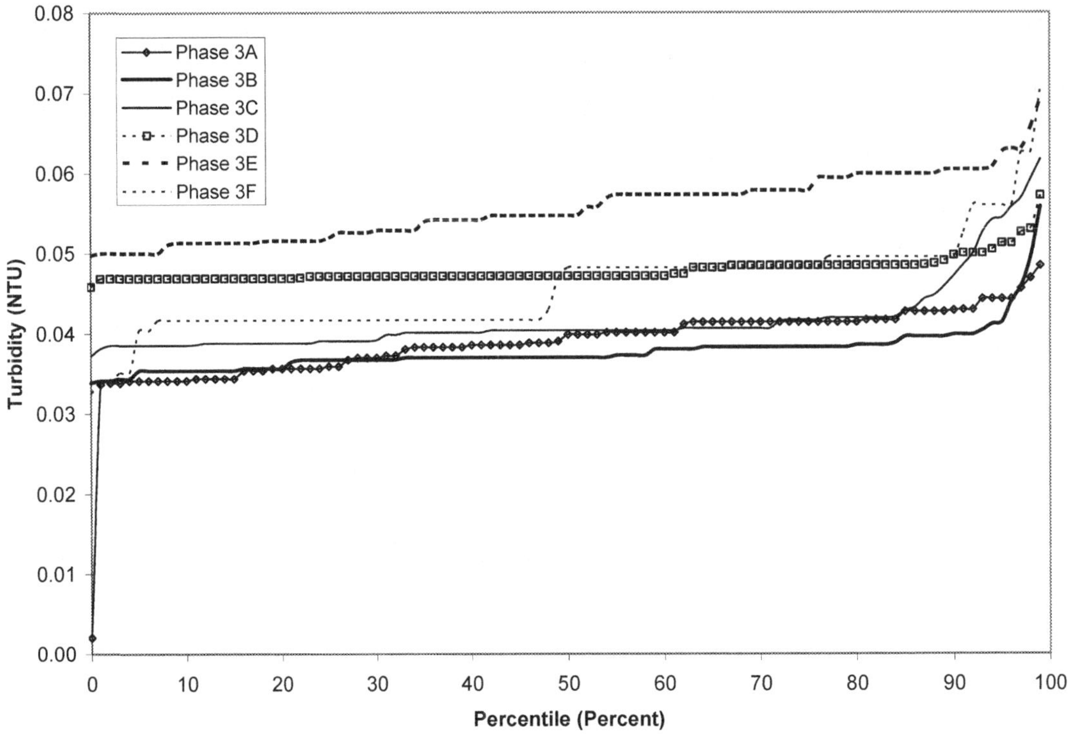

Figure 7.19 Turbidity percent less than plot – Filter 1

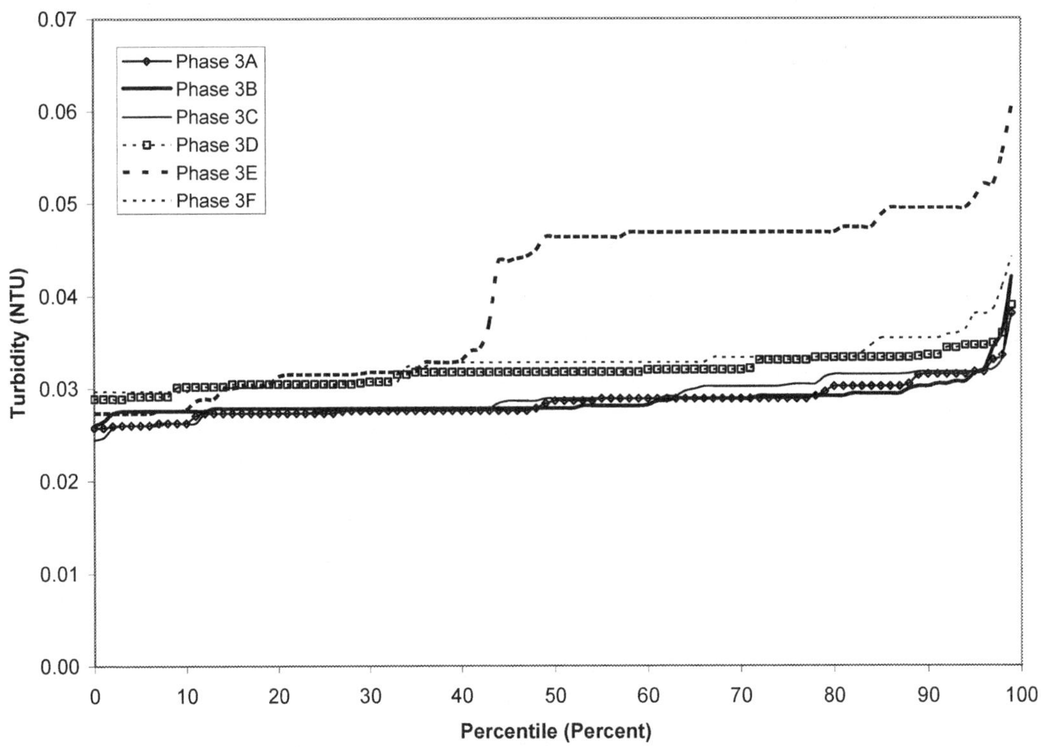

Figure 7.20 Turbidity percent less than plot – Filter 2

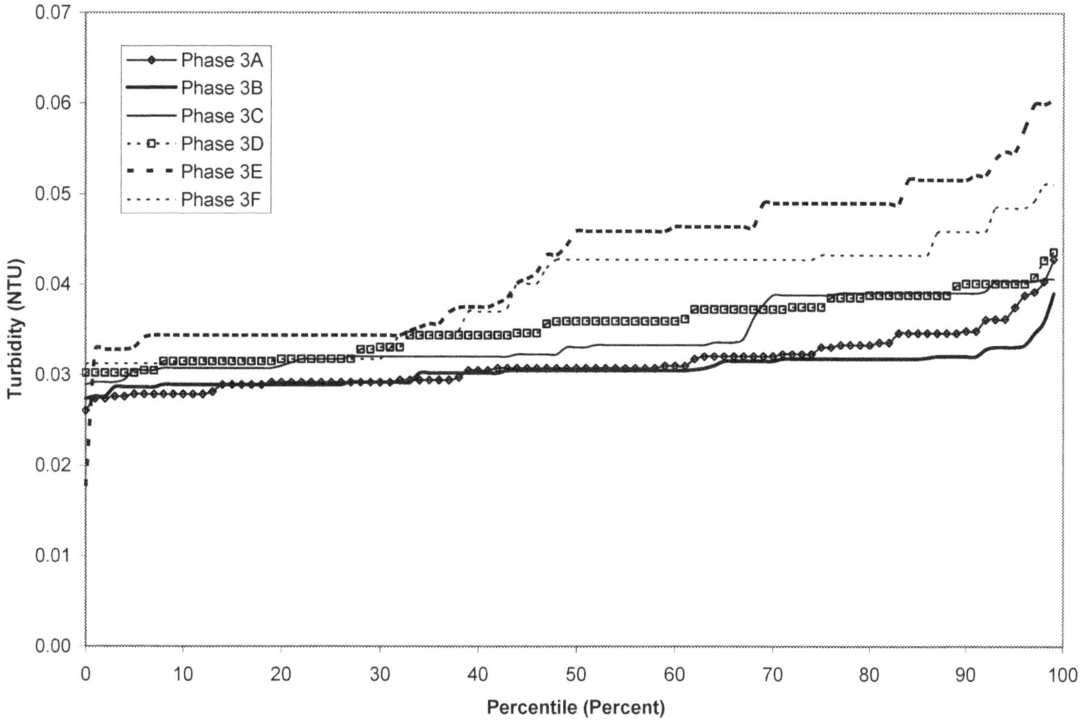

Figure 7.21 Turbidity percent less than plot – Filter 3

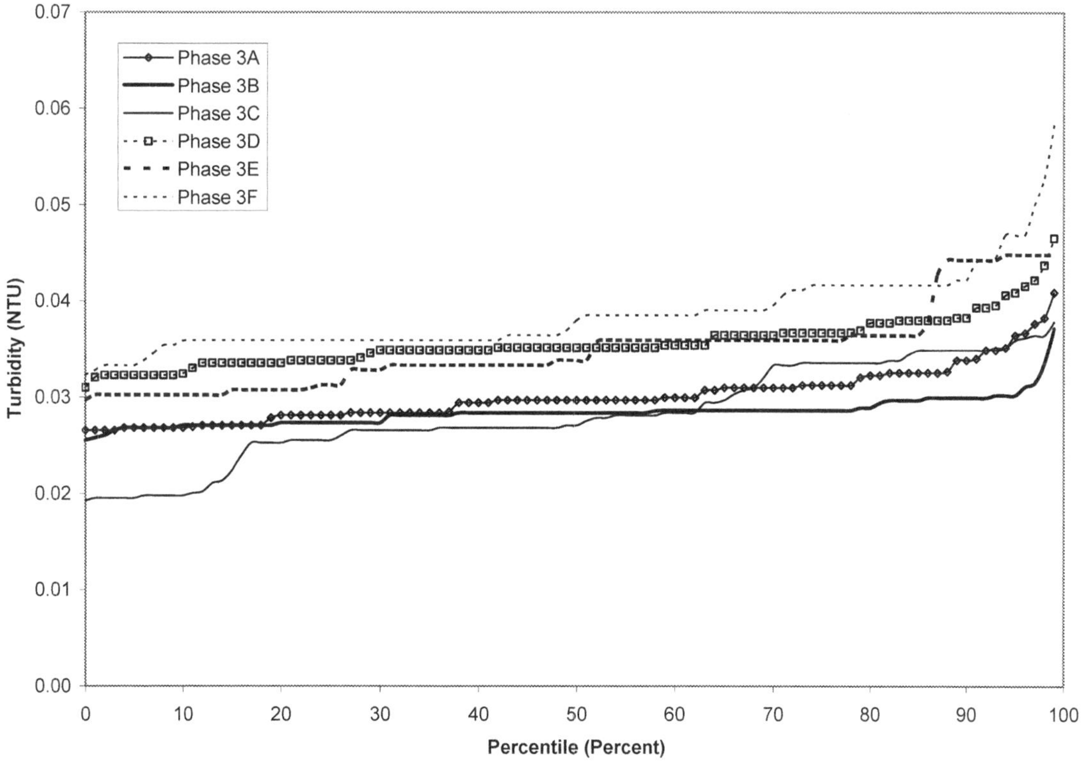

Figure 7.22 Turbidity percent less than plot – Filter 4

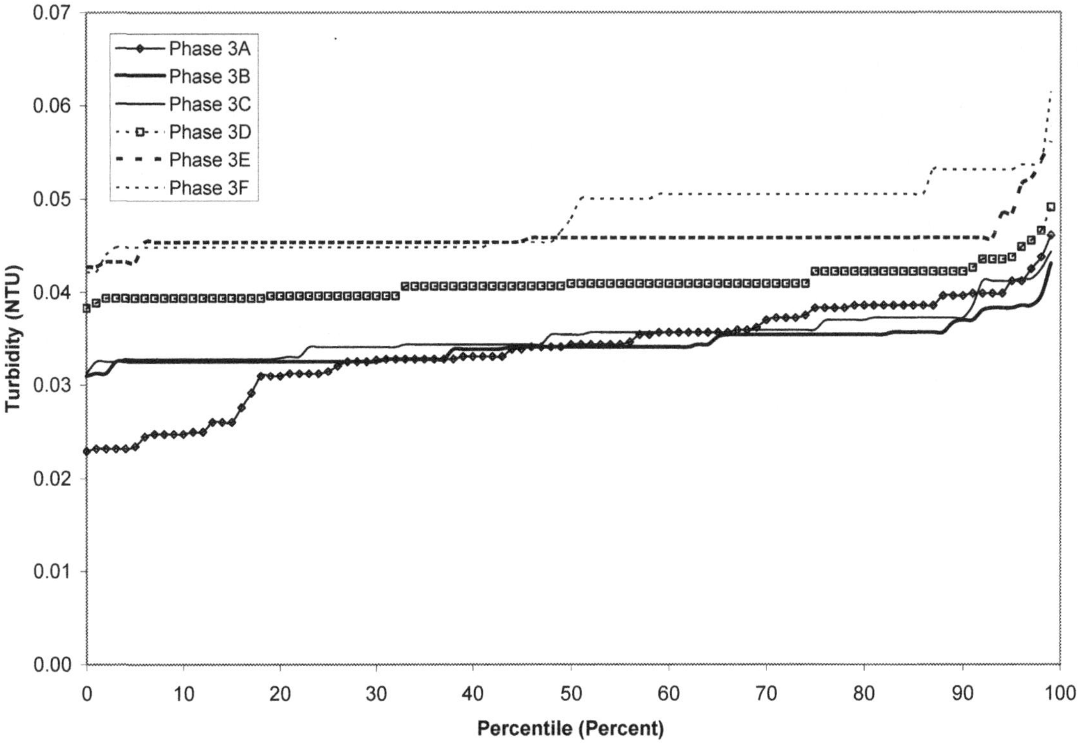

Figure 7.23 Turbidity percent less than plot – Filter 5

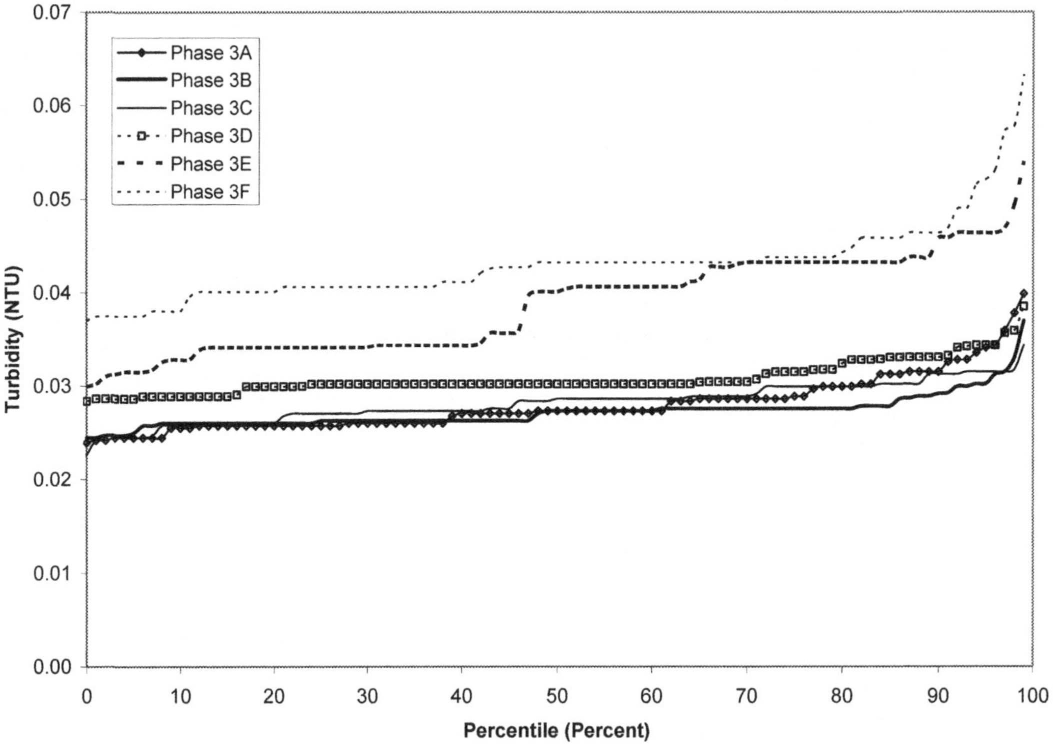

Figure 7.24 Turbidity percent less than plot – Filter 6

As shown in the preceding tables and figures, the filter effluent turbidities during Phases 3A, 3B, and 3C were all very similar. The 50th percentile turbidity values in any given individual filter effluent varied by less than 0.003 NTU during the three phases, and the 95th percentile values varied by less than 0.007 NTU. Excluding the combined filter effluent, the figures also show that the percentile rankings for all three phases followed the same trend, especially between 10 and 90 percent. Therefore, since the raw water quality was consistent and the coagulant and polymer dosages did not change substantially, according to the data from Phases 3A through 3C, substituting ClO_2 for a part or all of the preozonation dose did not increase the filter effluent turbidity.

Comparison of data for Phases 3D, 3E, and 3F, however, shows notable differences in the filter effluent quality. Generally the effluent turbidity values for Phase 3D were similar to those from the first three phases, but the values for Phases 3E and 3F were higher. The increase is more than likely a result of substantial changes in the raw water quality rather than changes in the preoxidation scheme. As has been discussed, during the last two phases the raw water composition changed substantially. Turbidity and organics concentrations increased, making it necessary to increase the coagulant and polymer dosages, which could have contributed to the higher filter effluent turbidities. Although the turbidities during these phases were higher than the other phases, all the 95th percentile turbidities were still well below 0.1 NTU.

Despite the fact that filter effluent turbidities were somewhat higher during Phase 3E and 3F, ClO_2 can be substituted for all or part of the preozonation dose without detriment to turbidity removal efficiency, even during times when raw water quality is poor.

TOC Removal

TOC removal was measured during testing to determine if substituting ClO_2 for ozone would change the organics removal during treatment. The average TOC removal data from the testing indicated that removal was not substantially affected by the preoxidation scheme. During Phases 3A, 3B, and 3C, the average TOC removal varied from 28 percent to 35 percent. The TOC removal increased during the last phase of testing, but the increase was likely due to the increase in raw water TOC and the change in the nature of the organics, not the change in preoxidation conditions.

During testing, the TOC concentration in the raw water was measured daily (Figure 7.6) and in the finished water weekly. A summary of the test conditions, raw water TOC concentration, finished water concentration, and percent removal for the long-term testing is listed in Table 7.6.

Table 7.6
TOC removal during long-term testing

Date	ClO_2 Dose (mg/L)	O_3 Dose (mg/L)	Raw Water TOC (mg/L)	Finished Water TOC (mg/L)	Percent Removal	Average Percent Removal
8/28/01	1.5	-	2.62	1.54	41	
9/4/01	1.5	-	2.58	1.75	32	35
9/11/01	1.5	-	2.68	1.86	31	
9/18/01	1.0	0.5	2.60	1.82	30	28
9/25/01	1.0	0.5	2.48	1.82	27	
10/16/01	-	1.0	2.58	1.65	36	
10/23/01	-	1.0	2.48	1.64	34	35
10/30/01	-	1.0	2.67	1.72	36	
11/13/01	1.5	-	2.50	1.66	34	34
11/20/01	1.5	-	2.53	1.64	35	
12/4/01	1.0	0.5	2.70	1.76	35	
12/11/01	1.0	0.5	3.33	2.07	38	37
12/18/01	1.0	0.5	3.51	2.21	37	
12/25/01	-	1.0	3.65	2.26	38	
1/1/02	-	1.0	4.60	1.99	57	51
1/7/02	-	1.0	5.68	2.45	57	

SDS Testing

The effect of varying the preoxidation conditions on DBP formation was evaluated by conducting weekly simulated distribution system (SDS) tests. Although CCWD uses chloramines for secondary disinfection and experiences no DBP formation problems, tests were conducted with free chlorine to generate data that would be useful to other utilities. The results showed that ClO_2 alone, a combination of ClO_2 and ozone, and ozone alone reduced the raw water DBP formation potential by about the same amount. The average THM reduction for each phase of testing was 35 percent to 45 percent, while average HAA formation was reduced 50 percent to 60 percent.

Details regarding the materials and methods used for determining SDS potentials are presented in Chapter 3, but the method is briefly summarized here. Samples of raw water and filter effluent were collected on a periodic basis. Finished water samples were collected at the filter effluent, because ammonia is added immediately ahead of the postozonation contactor for subsequent chloramination. The testing protocol called for weekly sampling; but sampling was less frequent because of laboratory and operating constraints. A summary of the sampling dates, testing conditions, and influent and effluent SDS THMs and SDS HAAs is presented in Table 7.7.

Table 7.7
DBP formation during long-term testing

Date	ClO_2 Dose (mg/L)	O_3 Dose (mg/L)	Raw Water SDS TTHM (ppb)	Filter Effluent SDS TTHM (ppb)	Raw Water SDS HAA5 (ppb)	Filter Effluent SDS HAA5 (ppb)
8/28/01	1.5	-	177	94	41	23
9/4/01	1.5	-	152	77	44	16
9/11/01	1.5	-	140	81	53	16
9/18/01	1.0	0.5	146	92	43	16
9/25/01	1.0	0.5	148	91	43	21
10/16/01	-	1.0	184	84	32	16
10/23/01	-	1.0	167	101	40	15
10/30/01	-	1.0	183	115	42	20
11/2/01	-	1.0	168	92	43	25
12/4/01	1.0	0.5	99	79	44	21
12/11/01	1.0	0.5	223	134	48	53
12/18/01	1.0	0.5	201	117	94	41
1/7/02	-	1.0	210	134	99	39

As shown in Table 7.7, the influent SDS DBPs varied widely from week to week and from phase to phase. Raw water THM formation varied from 140 to 223 ppb while HAA formation varied from 32 to 99 ppb. The variability of the DBP formation and the sampling make it difficult to draw definite conclusions about the effect of changing the preoxidation conditions during each phase. However, when the average percent removal for each phase is calculated, the results are comparable. The average DBP reduction for all phases was between 35 and 45 percent for SDS THMs and between 50 and 60 percent for SDS HAAs. So, substituting ClO_2 for ozone did not appear to substantially increase or decrease the DBP formation potential in comparison to ozone alone. However, additional testing would be needed to identify definite trends.

Concentrations of the four individual THMs that comprised the total THM concentration during the long-term testing were also compared. The percentages of the total THM formation accounted by chloroform (CF), dichlorobromomethane (DCBM), chlorodibromomethane (CDBM), and bromoform (BF) were calculated for each of the SDS tests. The average percentages of the total formation in the raw water for each phase are shown on Figure 7.25, and the averages in the filter effluent are shown on Figure 7.26.

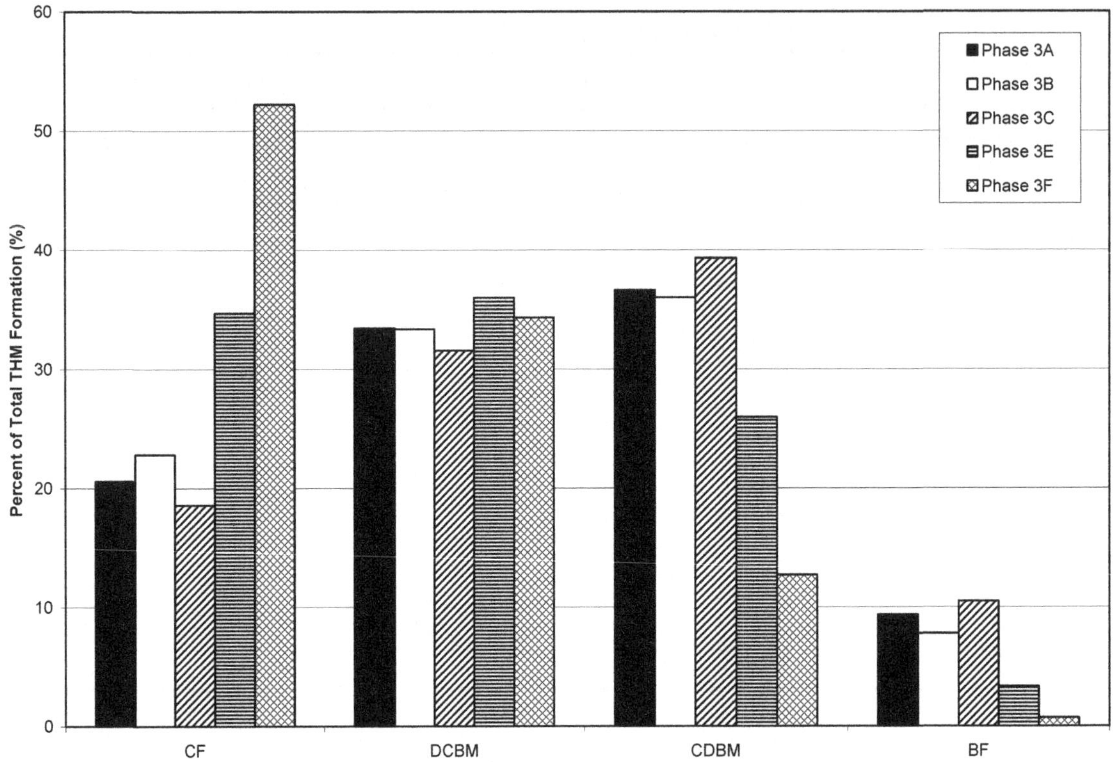

Figure 7.25 Average DBP speciation of raw water THMs determined by SDS potential tests

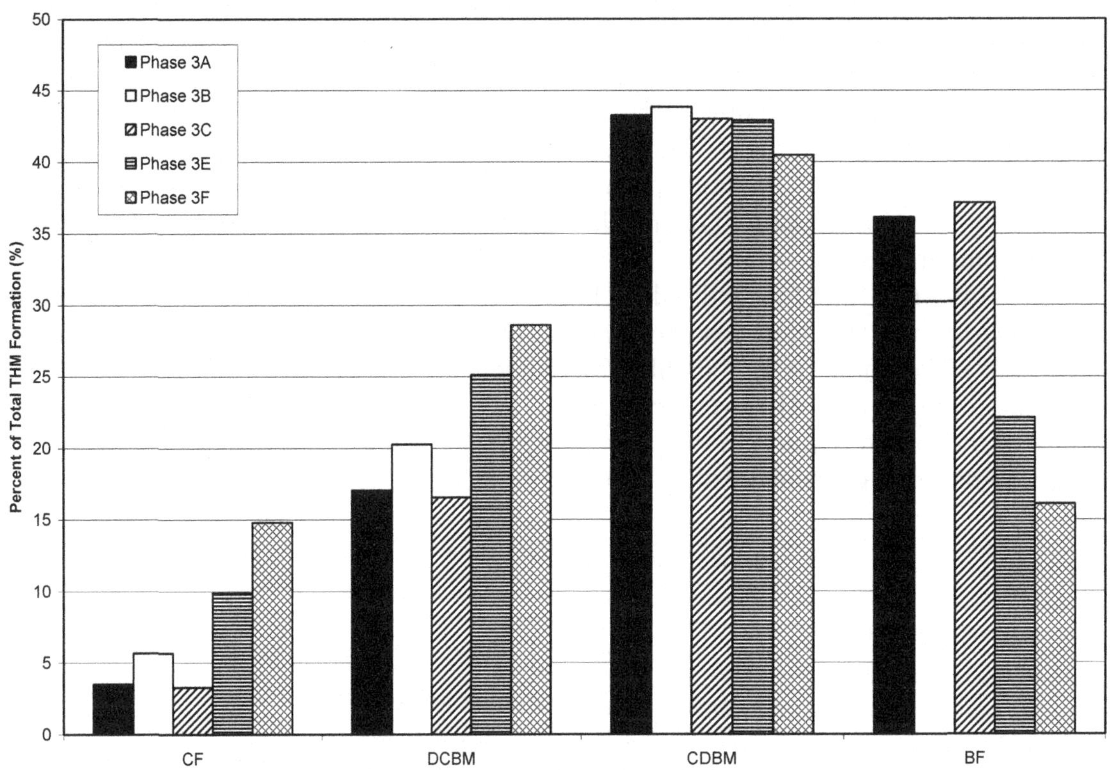

Figure 7.26 Average DBP speciation of filter effluent THMs determined by SDS potential tests

Generally, the raw water contained primarily chlorinated THMs, with a smaller fraction of brominated THMs; the reverse was true for the filter effluent. When the formations in the filter effluent are compared for Phases 3A, 3B, and 3C, there is little difference in the percent speciation. The results for Phases 3E and 3F differ slightly, which may be attributable to the changes in the raw water composition at the end of testing, as explained earlier. So, although preoxidation conditions were changed between phases, the change did not appear to substantially influence the distribution of chlorinated and brominated THMs in the filter effluent SDS.

OPERATIONAL ISSUES

An important consideration related to the use of ClO_2 is issues that affect daily operations; several were identified during the long-term testing program. One issue was that the ClO_2 required additional daily attention for the operators to tune the generator, perform process residual measurements, and the daily plant effluent compliance sampling and analysis. Another issue was adding ClO_2 requires additional monthly distribution sampling for chlorite. Also, because of the interaction of ClO_2 and chlorite with ozone, the operators had to monitor postozonation residuals more closely.

Adding a ClO_2 system to a plant that already uses ozone increased the daily operational requirements. One of the main reasons for the increase is that the amperometric titration method for ClO_2 analysis can be very time consuming, taking around 30 minutes to complete. The generator concentration and efficiency were checked daily using the amperometric titration method. The analysis of the generator took close to one hour to complete. During the special testing in July 2002, the electrochemical generator was used to produce ClO_2. Operation of this generator has shown the daily generator check could be reduced to a weekly check. Also, some generator manufacturers make a fully automated generator that requires less operator attention because it is equipped with an optical analyzer to monitor the ClO_2 feed solution concentration and adjust the generator accordingly. The plant effluent also had to be sampled daily for ClO_2 and chlorite concentrations using the amperometric titration method. The operators saved some time during this project by using the LGB kit for non-compliance ClO_2 residual measurements at different points in the process.

Monthly chlorite concentration compliance samples from the distribution system were required since ClO_2 was being added. This sampling was more complex than normal because of the somewhat intermittent use of ClO_2 and assuring sampling in the system corresponded to times when ClO_2 was being added or had recently been added. If ClO_2 is implemented permanently on the full-scale, the complexities will diminish and the sampling will become part of the routine monitoring program.

Adding a combination of ClO_2 and ozone required the operators to monitor the ozone process more closely. Since adding ClO_2 forms chlorite, and chlorite exerts an ozone demand as it is converted to chlorate, the ClO_2 dose can impact the ozone residuals. Reducing the preozonation dose could have also impacted the postozonation dose, since ozone is a stronger oxidant than ClO_2. Constant attention to the postozonation residuals was critical when the preoxidation conditions were varied because the postozonation dose needed to maintain adequate disinfection credit was affected (Table 7.2). The highest ozone doses were required when only

ClO_2 was added as a preoxidant or when ClO_2 was used in combination with ozone. It appeared that the higher postozonation doses were a function of the chlorite concentration after filtration. Therefore, if ozone is used for primary disinfection, care must be taken to monitor the ozone process and prevent ClO_2 overfeeding which would reduce the ozone residuals.

CONCLUSIONS

- Preoxidation with ClO_2 at the Randall-Bold WTP can be expected to drastically reduce bromate formation during preozonation and postozonation. In addition, preoxidation with ClO_2 can be expected to reduce the overall ozone use (preozonation plus postozonation).
- The chlorite that forms after ClO_2 will either be removed by GAC or oxidized to chlorate by ozone, so that the finished water chlorite concentration is below the detection limit.
- At the Randall-Bold WTP, chlorate will be the primary byproduct of ClO_2 preoxidation and the California AL for chlorate will limit the amount of ClO_2 that can be added.
- Using ClO_2 alone or in combination with ozone at the Randall-Bold WTP will not substantially change the chemical coagulation doses or increase the filter effluent turbidity levels.
- Preoxidation with ClO_2 or a combination of ClO_2 and ozone did not alter TOC and DBP formation potential removal when compared to using ozone only.
- Addition of ClO_2 into the treatment process at the Randall-Bold WTP will increase the sampling and analytical burden on the operations and laboratory staff and increase the complexity of operation because the operators will have to pay greater attention to ozone residuals in the postozonation contactor.

CHAPTER 8
BROMATE MITIGATION TECHNIQUE COMPARISON

Bromate mitigation techniques and their applicability to the Randall-Bold WTP were compared during special testing conducted in January 2002 and again in July 2002. Three techniques were investigated: preoxidation with ClO_2, pH adjustment, and ammonia addition. Details of the testing protocol and sample analyses are presented in Chapter 4.

BROMATE FORMATION

Results from the special testing showed that preoxidation with ClO_2 and pH adjustment were very effective in reducing the bromate formation. Ammonia addition had an effect, but to a much lesser extent. The July tests showed that using ozone alone would result in bromate concentrations exceeding the MCL, but either preoxidation with 1.0 mg/L ClO_2 or lowering the pH value to 6.0 reduced bromate concentrations to half the MCL. However, when the ozone dose was increased to 4.0 mg/L, the effectiveness of preoxidation with ClO_2 was decreased. The results also showed that substituting sodium chlorite ($NaClO_2$) for ClO_2 addition also reduced bromate formation, which strongly indicates that chlorite plays an important role in reducing bromate formation when ClO_2 is added.

Following completion of the Phase 3 testing, additional testing was conducted in January 2002 to evaluate the bromate formation reduction using preoxidation with ClO_2, pH adjustment, and ammonia addition. A summary of the conditions tested and the bromide, bromate, chlorite, and chlorate concentrations are listed in Table 8.1.

Table 8.1
Average bromate, chlorate, and chlorite formation in preozonation contactor during January mitigation testing

Condition*	Bromate (ppb)	Chlorate (ppb)	Chlorite (ppb)
Ozone only	3.2	<20	<10
0.5 mg/L ClO_2 preoxidation	<1	308	25
1.0 mg/L ClO_2 preoxidation	<1	303	30
pH adjustment to 6.5 units‡	2.2	17	<10
pH adjustment to 6.0 units‡	1.2	13	<10
0.44 mg/L NH_3-N residual	<5†	<10	<40
0.76 mg/L NH_3-N residual	<5†	<10	<10

*Preozone dose was held constant at 2 mg/L for all test conditions, average bromide concentration for all tests was 187 ppb

†Samples were sent to an outside laboratory with a different detection limit

‡Sulfuric acid dosages were 41 mg/L and 59 mg/L to reduce the pH to 6.5 and 6.0, respectively

During the January tests, the average bromate formation when using ozone alone was low (3.2 ppb) because of changes in the raw water quality, such as increased TOC. With the bromate formation close to the detection limit, it was difficult to compare the effectiveness of the mitigation techniques; however, some trends in the data are evident. The data show that the tests with ClO_2 addition resulted in an average bromate formation less than the detection limit while the tests with pH adjustment were slightly less effective in reducing bromate formation. No conclusions could be drawn about the effectiveness of ammonia addition because of the low bromate formation and because of the need to send the samples to a contract laboratory that had a different detection limit. Since the results from the mitigation comparison were not clearly defined, the testing was repeated during July 2002.

Also during January, special tests were conducted at higher ozone doses to determine the impact of ClO_2 preoxidation on bromate formation, ozone demand, and ozone decay. A summary of the ClO_2 and ozone doses and the bromate, chlorite, and chlorate concentrations are listed in Table 8.2. The average ozone residuals, ozone half-life, and integrated CT from the same tests are listed in Table 8.3. The integrated CT was calculated from the cell 2 residual, the decay rate, and the contact time.

Table 8.2
Average bromate formation in the preozonation contactor during ozone demand and decay testing

ClO_2 Dose (mg/L)	Ozone Dose (mg/L)	Average Bromate* (ppb)	Average Chlorite (ppb)	Average Chlorate (ppb)
0	3.6	9.2	<40	14
0.25	3.6	5.5	<40	130
0.5	3.6	<5	<40	260
1.0	3.6	<5	<40	390

*Samples were sent to a contract laboratory that had a different detection limit

Table 8.3
Average ozone residuals, half-life, and integrated CT in the preozonation contactor during ozone demand and decay testing

ClO_2 dose (mg/L)	O_3 dose (mg/L)	Average cell 2 residual (mg/L)	Average cell 3 residual (mg/L)	Half-life (min)	Integrated CT (mg•min/L)
0	3.6	0.19	0.06	2.0	0.30
0.25	3.6	0.30	0.02	0.9	0.24
0.5	3.6	0.31	0.04	1.1	0.31
1.0	3.6	0.42	0.02	0.7	0.28

As shown in Table 8.2, preoxidation with ClO_2 was effective in reducing the bromate formation, even with dosages as small as 0.25 mg/L. Although the detection limit for these tests was 5 ppb, a definite downward trend in bromate formation was observed as the ClO_2 dosage was increased. The bromate formation results from Table 8.2 contradict the results from the screening tests (Chapter 5), where 0.25 mg/L of ClO_2 actually increased the bromate formation. However, this study produced limited data on the effect of low doses of ClO_2 making it difficult to determine the exact impact on bromate formation. Therefore, the impact of low ClO_2 doses should be investigated further in the future. The results also showed that a majority of the chlorite formed during the ClO_2 addition was oxidized to chlorate by the ozonation.

The data in Table 8.3 show that as the ClO_2 dose was increased, the initial ozone residual (at the end of cell 2) increased. These data suggest that ClO_2 reduced the ozone demand of the raw water, which resulted in increased ozone residuals when the ozone dose was held constant. The decrease in ozone demand occurred despite the fact that ClO_2 is reduced to chlorite, which in itself exerts an ozone demand when it is oxidized to chlorate. Although the initial ozone residuals increased, the ozone residuals at the end of cell 3 did not increase and actually appear to have decreased slightly. However, the accuracy of ozone residual measurements at concentrations of about 0.05 mg/L or lower make it difficult to draw a definite conclusion. The ozone residual data were then used to calculate an integrated CT as an indication of the level of disinfection. The results show that although ClO_2 addition resulted in higher initial residual, the CT remained at a value similar to that observed when no ClO_2 was added. Although ClO_2 addition did not appear to increase the CT, the bromate formation decreased substantially when ClO_2 was added. The data regarding ozone demand and residual concentration were derived during a single series of full-scale tests and should be verified under better controlled conditions in bench-scale studies.

Since some of the results from the January testing were inconclusive because of low bromate formation and different detection limits of the analytical method used, similar tests were conducted over a period of three days during July 2002. These tests investigated bromate formation at two ozone doses (2.0 and 4.0 mg/L) and using multiple mitigation techniques such as preoxidation with ClO_2, sodium chlorite addition, pH adjustment, and ammonia addition. The addition of sodium chlorite as the source of chlorite provided an opportunity to evaluate whether ozone was reacting with chlorite and thereby limiting bromate formation or whether ClO_2 is the critical component that in some way interferes with bromate formation. A summary of the test conditions, ozone dosages, and the bromide, TOC, UV254, average bromate, average chlorite, and average chlorate concentrations is presented in Table 8.4.

Table 8.4
Average bromate, chlorite, and chlorate formation in the preozonation contactor during July 2002 mitigation testing

Condition	Ozone Dose (mg/L)	Bromide (ppb)	TOC (mg/L)	UV254 (1/cm)	Average Bromate (ppb)	Average Chlorite Post ClO_2 (ppb)	Average Chlorite Pre-Rapid Mix (ppb)	Average Chlorate Post ClO_2 (ppb)	Average Chlorate Pre-Rapid Mix (ppb)
Ozone only	4	180	2.1	0.095	60.3				
0.5 mg/L ClO_2	4	180	2.1	0.095	62.3	290	0	100	400
1.0 mg/L ClO_2	4	160	2.0	0.085	57.7	510	0	100	770
0.7 mg/L Chlorite	4	180	2.0	0.085	51.3	700	0	40	720
Ozone only	2	160	2.0	0.085	24.7				
0.5 mg/L ClO_2	2	180	2.1	0.095	14.3	300	0	70	370
1.0 mg/L ClO_2	2	180	2.1	0.095	**5.2**	510	0	130	720
0.7 mg/L Chlorite	2	180	2.0	0.085	**5.3***	710*	0*	30*	630*
pH 6.5	2	180	2.0	0.085	13.3				
pH 6.0	2	180	2.0	0.085	**5.6**				
0.25 mg/L NH3-N	2	210	2.0	0.087	19.7				
0.5 mg/L NH3-N	2	210	2.0	0.087	16.9				

*Represents data from third sample, not the average of three samples

As shown in Table 8.4, when 2.0 mg/L of ozone alone was added, the bromate formation from the preozonation contactor averaged 24.7 ppb, well in excess of the MCL. When ClO_2 was added without changing the ozone dose, the bromate formation was reduced substantially. A ClO_2 dose of 1.0 mg/L reduced the average bromate formation to 5.2 ppb, which is almost an 80 percent reduction in bromate formation when compared to no ClO_2 addition. When 0.7 mg/L of chlorite was added ahead of the preozonation contactor, the bromate formation was reduced to 5.3 ppb, which is very similar to the effect of 1.0 mg/L of ClO_2. Therefore, the data suggest that the chlorite formed after ClO_2 addition, plays an important role in reducing bromate formation.

Testing was also conducted using pH adjustment and ammonia addition for bromate mitigation. The data showed that pH adjustment was very effective. Reducing the pH to 6.5 lowered the average bromate formation to 13.3 ppb, and reducing the pH to 6.0 lowered the average bromate formation to 5.6 ppb. Ammonia addition had an effect, but to a much lesser extent - an ammonia dosage of 0.5 mg/L reduced bromate formation only to 16.9 ppb.

The testing done with an ozone dose of 4.0 mg/L showed very high bromate formation. When ozone alone was added, the average bromate formation was 60.3 ppb, well above the MCL. When ClO_2 or chlorite were added to the raw water before adding ozone, the bromate formation was decreased, but not to the extent observed at lower ozone dosages. The 1.0 mg/L ClO_2 dose that reduced bromate formation by 80 percent at an ozone dose of 2.0 mg/L, reduced bromate formation only marginally with an ozone dose of 4 mg/L. A chlorite dose of 0.7 mg/L resulted in the lowest bromate formation, but only about a 15 percent lower than with ozone alone. Although ClO_2 preoxidation was less effective, an ozone dose of 4.0 mg/L is higher than the typical dose used at the Randall-Bold WTP and the results of a 2.0 mg/L ozone dose are more applicable.

Preoxidation with ClO_2 and pH adjustment appear to be the best methods of reducing the bromate formation at the Randall-Bold WTP with typical ozone doses. Both were very effective in reducing the bromate formation below the MCL, even when the use of ozone alone resulted in bromate formation of about 25 ppb. The data also suggested that chlorite addition is as effective as ClO_2 preoxidation in reducing bromate formation. Before any decision is made regarding substitution of chlorite for ClO_2, however, detailed bench-scale testing should be conducted to gain a better understanding of the complex reactions and the potential for byproduct formation. Also, adding sodium chlorite will not maintain the filtration effects associated with ClO_2 preoxidation that were observed during Phase 3.

PERCHLORATE FORMATION

During July 2002, additional samples were collected for analysis of perchlorate to determine whether chlorate could be oxidized to perchlorate by ozone. Samples were collected on July 24 from the raw water and the finished water. During that period, ClO_2 was being added at 0.5 mg/L, the preozonation dose was 1 mg/L, and the postozonation dose was 0.5 mg/L. Split samples were sent to the California DHS laboratory and to MWH laboratories. The results showed that the concentrations in both samples were below the detection limit (2.5 ppb for the samples sent to DHS and 4.0 ppb for the samples send to MWH). The absence of perchlorate formation following ozonation agrees with the findings of Hoigné et al. (1985), who stated that chlorate does not react with ozone.

TREATMENT CONSIDERATIONS

Besides effectively reducing the bromate formation during ozonation, each of the mitigation techniques investigated will have other treatment considerations that must be addressed. The following are some of the positive and negative aspects of each of the mitigation techniques.

Preoxidation with Chlorine Dioxide

Preoxidation with ClO_2 was very effective at reducing the bromate formation during ozonation; however, ClO_2 use probably affects the rest of the treatment plant more than either of the other two mitigation techniques. As discussed previously, after ClO_2 addition, chlorite is formed which exerts an ozone demand and required higher postozonation doses. Also, Gordon and Bubnis (1999) stated that chlorate formation is undesirable in drinking water and suggest that ozone should be added prior to ClO_2 to avoid its formation. The ClO_2 system is arguably the most complex of all the feed systems at the Randall-Bold WTP. Despite the negatives associated with ClO_2, the long-term testing showed that ozone could be substituted either for a portion or all the ozone dose without substantially impacting the filtration process. Another potential benefit, although regulatory credit is unlikely, is a potential disinfection synergy when two disinfectants are used in sequence. Anderson et al. (2001) showed that when ClO_2 disinfection was followed by chloramine disinfection, there was a disinfection synergy with *B. subtilis* spores. In addition, with two disinfectants used for preoxidation, if one is out of service, the production from the other can be increased to maintain preoxidation.

Adjustment of pH

During this study, pH adjustment to a value of 6.0 was effective at reducing bromate formation; however, the sulfuric acid dosages required to reach this level were very high. During the January testing it took 59 mg/L and during the July testing it took 49 mg/L. At these levels of acid addition, about 60-65 mg/L of caustic would be required to raise the finished water pH to its target value of 9. Both the sulfuric acid and caustic dosages are excessive and would add to the already high TDS concentration in the water. An issue with pH adjustment is lowering the pH will impact the coagulation and filtration processes. Although lowering the pH could increase the TOC removal, it may affect coagulation and create non optimal conditions for the filtration process. More testing would be needed to determine the impact, whether positive or negative.

Ammonia Addition

Ammonia addition would be the easiest bromate mitigation method to implement at Randall-Bold because an ammonia feed system is already in place; but it was not as effective as ClO_2 preoxidation and pH adjustment. The ammonia doses required during this study were relatively high, but Randall-Bold WTP has one advantage versus other facilities, it has to add ammonia for chloramine formation. One of the primary disadvantages of ammonia addition is that if free chlorine is used for disinfection at some point later in the process, chlorine must be added to breakpoint the ammonia before a free chlorine residual can be maintained. The ammonia dosages investigated during this study could require 5 to 7 mg/L of chlorine just to remove the ammonia. However, since Randall-Bold WTP uses chloramines (ammonia is added prior to chlorine) and it uses ozone for primary disinfection after filtration, there is no need to breakpoint the ammonia added for bromate mitigation. In fact, the Randall-Bold WTP in essence already practices ammonia control by adding the ammonia for chloramine formation prior to the postozonation contactor and then adding the chlorine after the postozonation contactor.

CONCLUSIONS

Conclusions from the bromate mitigation testing were:

- Preoxidation with as little as 0.25 mg/L of ClO_2 before preozonation can substantially reduce the bromate formation.
- Chlorite addition in the form of sodium chlorite can reduce bromate formation equal to ClO_2 addition, which demonstrates that the chlorite ion plays an important role in reducing the bromate formation. If sodium chlorite is substituted for ClO_2, other benefits of preoxidation, especially improved filter performance, will not be realized.
- Reducing the raw-water to pH 6.0-6.5 prior to preozonation can reduce bromate formation about as well as preoxidation with ClO_2; the benefits are greater at the lower pH.
- Ammonia addition to raw water at dosages of 0.25 mg/L and 0.5 mg/L can reduce bromate formation but not to the same extent as either ClO_2 preoxidation or raw-water pH reduction to pH 6.0.
- Perchlorate will not be formed by ozonation of water containing chlorate.
- Each of the bromate mitigation techniques investigated during this study have advantages and disadvantages of their application at the Randall-Bold WTP.

CHAPTER 9
ENERGY USES AND IMPACTS

The implications and the potential impact that combining the use of ClO_2 and ozonation has on water treatment energy use are discussed in this chapter. Two perspectives are considered: (1) the impact of this combination on the Randall-Bold WTP's energy use and (2) in light of the 2001 crisis in California's electrical supply system, the potential impact of implementing a similar combination in all ozone-using water treatment plants across the state. Because preliminary assessments conducted before this report suggested that a statewide implementation could result in a significant reduction in energy use, a general analysis based on the results of this study was conducted to verify that assessment.

EPRI/AWWARF OZONE EFFICIENCY STUDIES

The use of ozone has always been associated with high energy consumption. Given the many financial constraints that many water treatment systems face, various efforts have been made to reduce the amount of energy required to use ozone in water treatment. In 1996, AwwaRF and the Electric Power Research Institute (EPRI) launched a collaborative, three-phased study aimed at standardizing the practices used by water treatment plants in the generation of ozone, in order to lower their energy use and to speed their acceptance of ozone use. This effort led to the development of an assessment procedure. The procedure, which was refined by its use in over a dozen water treatment plants in North America, has proven effective at reducing energy use at ozone water treatment plants by 10 to 50 percent, depending on the plant. Energy use reductions were achieved through a combination of optimization methods, equipment maintenance, and development of specific goals for use by plant operators in daily planning and operation. This study was one of the many collaborative endeavors by the two organizations in the areas of water treatment practices influenced by both water quality and energy use. The results of these efforts served to familiarize electric utilities throughout North America with the unique needs of water treatment plants.

The optimization process focuses on the generation and application of ozone. More recent research within the drinking water industry has demonstrated water quality benefits when multiple disinfectants and oxidants are applied in tandem, but the possible energy savings derived from this practice have not been quantified. The high energy requirements associated with the use of ozone (in comparison with alternative disinfectants) continue to make ozone a target for optimization and cost-cutting measures. The current study and the attendant changes in energy use at the Randall-Bold WTP are a natural outgrowth of ozone efficiency optimization efforts.

ENERGY USES FOR TREATMENT SCHEMES IN THIS STUDY

During the testing period in late 2001 (from September through December), energy use rose and fell as ozone dosages were increased and decreased as would be expected, given the higher energy use necessary for generating ozone in comparison with ClO_2. However, when ClO_2 was added, the overall ozone use was not reduced at a 1:1 ratio, implying that the relationship between chemical use, water quality, and plant energy use is more complex than initially envisioned. The fact that ClO_2 addition forms chlorite and chlorite creates a demand in the postozonation contactor which minimizes the potential energy savings and has the net effect of limiting the amount of energy to be saved by ClO_2 and ozone. The daily, seasonal, and annual trends in energy are discussed in more detail below.

Daily, Seasonal and Annual Trends

The testing period extended from September 2001 through December 2001, as detailed in Chapter 4. The energy consumption trends associated with ozone and ClO_2 use at the Randall-Bold WTP during the testing period are presented on Figure 9.1. Data collected from the testing of the R-101 ClO_2, Sterling Pulp Chemical's more advanced ClO_2 generator, at the Randall-Bold WTP in the summer of 2002, are not included in this figure.

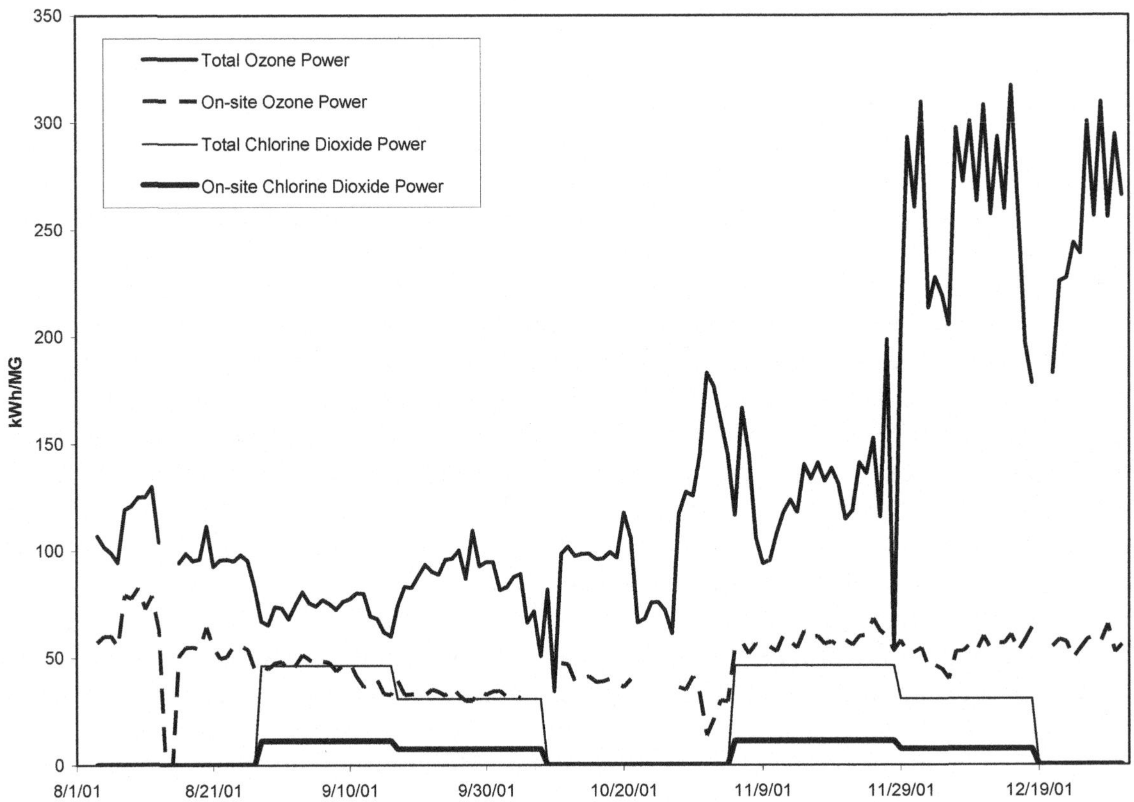

Figure 9.1 Power consumption for ozone and ClO_2 at Randall-Bold WTP

The figure presents both "on-site" and "total" power consumption for ozone and ClO_2 generation. The power consumption figures are normalized (i.e., expressed as kilowatt-hours per million gallons) in order to eliminate the effect of changes in treated water flows on the ozone system power consumption. Thus, before the large "jump" observed in total ozone system power use after mid-November, the fluctuations shown in total and onsite ozone power are largely a function of the change in ozone dose, as detailed in the testing protocol section in Chapter 4.

On-site vs. Total Power Use

In-plant electricity consumption is only a portion of the total energy use associated with producing either ozone or ClO_2. Ozone is generated from either air or oxygen, and requires electricity in order to be processed and stored. Chlorine dioxide requires sodium chlorite and chlorine which, in turn, are generated through electricity consuming processes. Given the California Energy Commission's interest in conserving statewide energy supplies, the total energy requirements of generating both of these chemicals as well as the potential impact of substituting ClO_2 for ozone were analyzed. Ozone is generated from liquid oxygen, which is produced cryogenically, or from compressed air, which is generated on-site. As was expected, the data indicates that both the power consumption at the plant and the total power use are significantly greater for ozone than for ClO_2, reflecting the energy intensive nature of the ozone generation process. In fact, energy consumption associated with ozone use is often as much as five times the energy consumption needed to conventionally generate ClO_2. Energy requirements associated with generating ozone and generating ClO_2 are compared in Table 9.1.

Table 9.1
Comparison of on-site and off-site power requirements for ozone and ClO_2

Chemical	On-site	Off-site Power
Ozone*	Generation • 8 kWh/lb from air • 5 kWh/lb from oxygen Gas Preparation • 3 kWh/lb to compress and dry air • 1.8-2.0 kWh/lb ozone to generate gaseous oxygen on-site, assuming 10 lb oxygen for every pound of ozone. Contacting • Wide range depending on type of contactor. Porous bubble diffusers require no additional energy while high pressure partial injection requires as much as 20 kWh/lb ozone. Off-gas Destruction • 2 kWh/lb ozone	Liquid Oxygen Generation • 1.3-1.7 kWh/ lb ozone, assuming 10 lbs oxygen needed for 1 lb. ozone.
ClO_2	Generation • 0.9 kWh/lb from sodium chlorite and chlorine gas	Sodium Chlorite Generation • 4.3 kWh/lb from sodium chlorate, HCl, H_2O_2 and NaOH

*Ozone data adapted from Rice (1996)

Rise in Total Ozone Power Use

Because of a unique set of circumstances at the plant total power consumption for ozone generation rose significantly near the end of the testing period. Figure 9.1 shows that near the end of the testing period (around the middle of November) the energy consumption associated with ozone generation rose dramatically. The Randall-Bold plant uses liquid oxygen as the feed gas through the ozone generators. The generators need a minimum amount of gas passing through the system; otherwise, the equipment protection device will override the generators and shut them down. At the very low flow rates (less than 3 mgd, or 4 percent of total capacity) encountered near the end of the testing period, the ozone consumption was reduced correspondingly, but because of the safety overrides on the ozone generator, the liquid oxygen consumption could not be reduced below a certain minimum. Thus, the relative rate of liquid oxygen consumption (i.e., the amount per million gallons of water treated) rose substantially, driving up the normalized total power consumption.

Except for a short time at the end of the testing period, the day-to-day power consumption associated with ozone use seems to vary only slightly. This data reflects that, while the preozonation dose ranged between 0 and 1.5 mg/L, the total ozone dose hovered mostly within the range of about 1.2 to 1.7 mg/L. The postozonation system is operated to maintain an ozone residual in the contact basin for disinfection. Given that the ozone residual can be produced only after the overall ozone demand has been met, a 'minimum' ozone dose is necessary to meet this demand. Unfortunately, since ClO_2 reduces the ozone use by only a slight amount (about 0.2 mg/L per 1.0 mg/L of ClO_2), its impact on the overall ozone use and the associated energy needs is minimal.

The test was conducted over a four month period, which included the part of the year when the plant's raw water quality is typically at its poorest. The poor water quality is reflected in high ozone demand; thus, the results can be used to produce a conservative estimate of the reduction in the plant's electrical demand by changing from preozonation to using ClO_2. Also, given the small reduction in the total ozone dose observed in this study, it can be concluded that overall electricity consumption would be reduced only slightly by switching preoxidants. Based on an average plant flow of 28 mgd and a reduction in ozone use of 0.3 mg/L, the annual savings in electricity use would be approximately 104,000 kWh. On the other hand, there is at least circumstantial evidence that the efforts of plant staff to employ ozone optimization techniques at the plant yielded considerable benefits. The effects of the optimization efforts on the Randall-Bold plant, and their implications for statewide energy use, are discussed below.

Study Implications at Randall-Bold Plant and Statewide

The power requirements for ozone generation at the Randall-Bold plant are a function of ozone dose, which in turn is a function of ozone demand. Ozone demand is a function of raw water quality and, in the case of this plant, ozone demand was created by the formation of chlorite after ClO_2 addition. Thus, ClO_2 preoxidation replacing some preozonation, only has a minor impact on reducing power consumption. In fact, according to the data collected during the testing period, the reduction in power demand at the treatment plant would range from about 20 kW at average flow to approximately 85 kW at peak flow. These estimates are based on the assumption that operators would apply the highest dose at peak flow, which may no longer be true, given their recent efforts at optimization. Thus, actual power demand reductions will probably be even smaller.

Given the unique raw water quality at the Randall-Bold plant, the data generated during this study must be used with caution when estimating the statewide impacts of changing oxidants. The plant treats water which is tidally influenced from the San Francisco Bay. The quality of the raw water received in California's water treatment plants varies immensely. The Randall-Bold WTP is also unique in its use of ozonation after filtration. It is likely that many plants that currently use ozone would achieve a greater reduction in the overall ozone dose by using ClO_2 as a preoxidant than achieved at the Randall-Bold plant.

More important, recent research indicates that using a combination of disinfectants is more effective at inactivating certain pathogens and other microorganisms than reliance on a single one. Thus, using ClO_2 ahead of ozone in water treatment will have benefits beyond the reduction in power consumption. However, if replacement ratios similar to those used at the Randall-Bold plant were used by all water treatment plants in California that use ozone (a maximum of 2,575 mgd), a conservative estimate of the statewide demand reduction would be around 2.4 MW. Additional power demand reductions resulting from eliminating the production of liquid oxygen that is no longer needed as feedstock would raise the statewide reduction in power demand to about 4.5 MW.

Power Consumption Comparison with Past Operation

Operating data from the Randall-Bold water treatment plant were calculated in an effort to make a comparison between the plant's energy consumption under past operating practices and during the testing period. Such a comparison is crucial when assessing the potential impact of the proposed change in oxidants on the plant's overall power consumption. Average monthly power consumption by the plant's ozone system for the period of August 2000 through December 2001 is plotted on Figure 9.2.

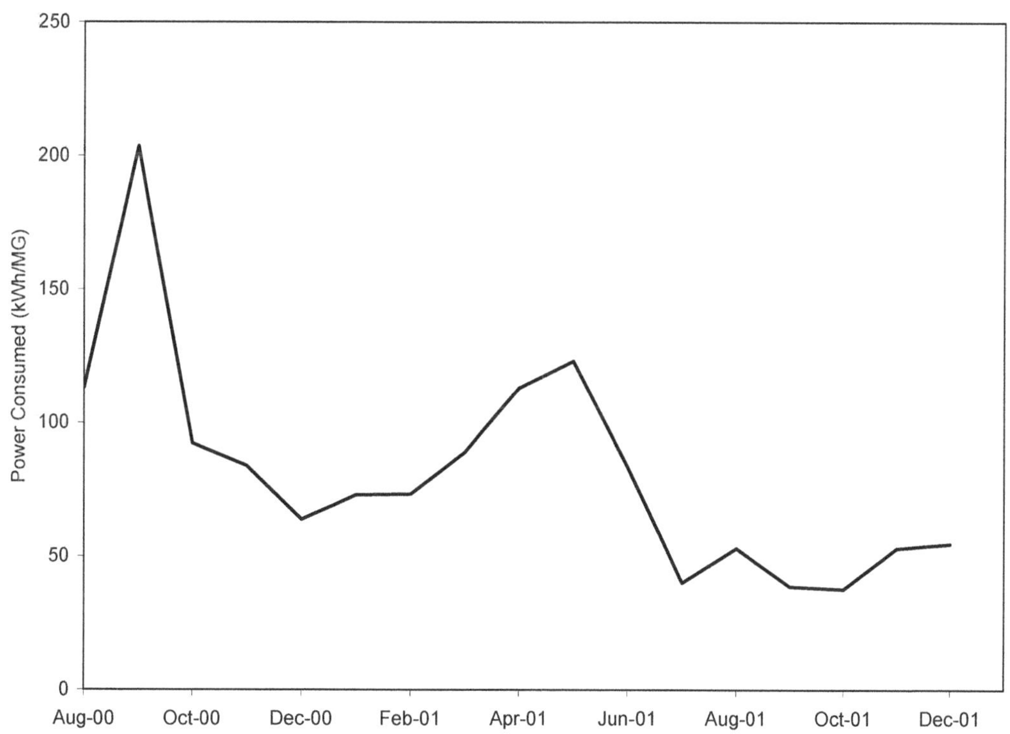

Figure 9.2 Power consumption for ozone at the Randall-Bold WTP from August, 2000 through December, 2001

Power consumption in the months preceding the testing ranged between 60 and 200 kWh/MG largely because of the variation in the applied ozone dose. By contrast, the average power consumption throughout the testing periods from September through December 2001, was fairly constant at approximately 50 kWh/MG, even though the testing was conducted during a period of poor raw water quality, when ozone demand is usually expected to be high.

Two factors account for this trend. The first factor, mentioned earlier, was the 'base' ozone demand exerted by the water. So, while the preozonation dose ranged from 0 to 1.5 mg/L, the total ozone dose during the testing period typically ranged between 1 and 2 mg/L, and averaged 1.5 mg/L. The second factor was the realization by the plant staff before the testing began, that the ozone dosages used at the plant may have been excessive. After a thorough investigation and a process evaluation, the average ozone dose was lowered from 2.2 mg/L to about 1.5 mg/L, which was found to meet the treatment objectives of taste and odor removal and disinfection. The procedure used to adjust the dose is similar to the optimization process developed by EPRI/AWWARF. In fact, this achievement is further evidence of the beneficial effects which can be attained through the efficiency process.

Evaluation of the data on power consumption before the testing indicates that replacing ozone with ClO_2 in the preoxidation basin will have only the minor impact on the overall power demand. However, before the testing began, it was discovered that the average ozone dose used at the Randall-Bold WTP was excessive and could be safely reduced, which would result in a corresponding reduction in the energy plant's consumption. The situation has inadvertently demonstrated that optimization of the ozonation process may yield bigger energy savings than replacement of oxidants. In the case of the Randall-Bold WTP, the unique chemistry of the raw water is the primary reason for switching to the use of ClO_2 as a pre-oxidant. The lesson learned from this study is that other water treatment plants in California that treat other waters of different composition should not base their decision to oxidants on potential energy savings. If reducing energy consumption is their principal goal, they might be well advised to conduct a thorough evaluation of their ozonation system's operation.

Given the results of the energy evaluation and the reduction in energy use achieved through improving the ozone generation and application processes, statewide adoption of alternative disinfectants would yield modest reductions in electricity demand in the state of California. Thus, a broad switchover would be warranted only if water quality concerns dictate. In the absence of compelling concerns about the effect of ozone on the bromide containing raw water, it would be more beneficial from an energy standpoint if the water agencies that use ozone assess their practices and adopt more energy-efficient measures wherever possible. For a state agency with limited resources, such as the California Energy Commission, promoting more energy efficient ozone generation practices may be a better use of funds than alternative disinfection schemes. Electricity demand and energy use could be greatly improved by such a program.

Comparison with Alternative Disinfection Technologies

Ozone is one of the most energy-intensive potable water disinfection technologies commonly found in U.S. water treatment plants. Thus, under virtually any scenario using a disinfectant to replace part or all of an ozone dose will result in less electrical consumption and a drop in electrical demand.

The most common alternative disinfectant is chlorine, which has a very low electrical impact. However, comparing the energy requirements of alternative disinfectants ignores the reasons behind implementing ozone in the first place. Ozone is a superior oxidant compared to chlorine and does not produce chlorinated disinfection byproducts, including trihalomethanes. Thus, it is very effective at treating a multitude of raw water problems, such as taste and odor or iron and manganese. So, for all of ozone's complexity and high energy use, it is often the best option when confronted with poor raw water quality, which is commonly encountered in the state of California and across the country.

Discussions on disinfection processes at water treatment plants should be driven by water quality issues rather than power consumption concerns. Modest reductions in energy use are possible by using a variety of disinfectants and oxidants, but the small reductions suggested by the results in this study point to the need to evaluate alternatives using other criteria. If research demonstrates the efficacy of multiple disinfectants in a water treatment process, various combinations should be pursued for the sake of improved water quality rather than gains in energy savings.

Conclusions

Conclusions from the energy uses and impact analysis were:

- Using ClO_2 in tandem with ozone will produce a modest reduction in the energy demand and electricity consumption at water treatment plants. Given the benefits of water quality improvement, moderate energy saving and increased complexity associated with using multiple chemicals, planners should weigh water quality impacts more heavily than energy savings.
- The optimization of ozone generation and use can substantially reduce energy use.
- From the standpoint of California Energy Commission, who is committed to lowering statewide energy demand, the projected savings of replacing preozonation with chlorine dioxide (based on the results of this study and extrapolated to the entire state), are moderate. If energy demand reduction is the principal goal, the California Energy Commission should not commit extensive resources to implementing this change across the state. On the other hand, with potential water quality benefits of combining the two disinfectants/oxidants, the benefit of moderate energy saving is certainly an added bonus.

CHAPTER 10
CONCLUSIONS AND RECOMMENDATIONS

CONCLUSIONS

The principal research objective of the project, stated broadly, was to evaluate the benefits of using ClO_2 in combination with ozone. The specific objectives for the evaluation and whether the project met the objective included the following:

Objective 1- Determine the extent to which ClO_2 preoxidation reduces the ozone demand/dose in the preozonation and postozonation contactors.

- This objective was met. Although preoxidation with ClO_2 forms chlorite, and the chlorite creates ozone demand for the postozonation contactor, the total ozone dose (preozonation plus postozonation) was reduced.

Objective 2 - Determine the extent to which ClO_2 preoxidation reduces the bromate formation by reducing the ozone dose.

- This objective was met. The data from this study showed that preoxidation with ClO_2 substantially reduced the bromate formation. In addition to reducing the bromate formation by reducing the ozone dose through substitution with ClO_2, an unexpected result was that ClO_2 preoxidation also reduced the bromate formation at a constant ozone dose. Special testing showed that chlorite, which is formed after ClO_2 addition, plays an important role in the bromate formation reduction.

Objective 3 - Determine the extent to which ClO_2 preoxidation reduces the energy consumption both at Randall-Bold WTP and statewide.

- This objective was met. Using ClO_2 in combination with ozone resulted in a modest reduction in the energy demand and electricity consumption at the Randall-Bold WTP; however, the projected savings (based on the results of this study and extrapolated to the entire state) are not large.

Objective 4 - Determine the extent to which ClO_2 preoxidation results in chlorate ion formation by ozone oxidation of chlorite ion.

- This objective was met. The data from this study showed that ozone oxidized the chlorite to chlorate, so that the plant effluent chlorite concentration was below the detection limit. However, because a majority of the chlorite is converted to chlorate, the California AL for chlorate (0.8 mg/L) will limit the ClO_2 dosage applied at Randall-Bold. The testing also showed that some of the chlorite was removed by the GAC, while chlorate neither increased nor decreased through the GAC filters.

Objective 5 - Determine the extent to which ClO_2 preoxidation maintains the filtration benefits of preozonation.

- This objective was met. Preoxidation with ClO_2 alone or ClO_2 in combination with ozone required similar coagulant dosages and produced similar filter effluent turbidity when compared to using ozone alone.

Objective 6 - Determine the extent to which ClO_2 preoxidation reduces THM and HAA formation potentials in finished water.

- This objective was met. The data showed that there was not a substantial difference in DBP formation potential between the three preoxidation conditions tested during the long-term phase of this study.

In addition to the conclusions that supported the objectives of this study, the data produced other general conclusions that included:

- Preoxidation with ClO_2 and pH adjustment to a value of 6.0 were effective methods to reduce the bromate formation in the preozonation contactor. Ammonia addition before the preozonation contactor was effective, but to a lesser extent than ClO_2 and pH adjustment.
- Implementing ClO_2 in the treatment process at Randall-Bold WTP added complexity to the operations staff that included additional sampling, time consuming amperometric analyses, and the need to pay close attention to the postozonation dose because of the ozone demand created by chlorite.
- Optimization of ozone generation and ozone dose can have a large impact on the energy uses at a water treatment plant.

RECOMMENDATIONS FOR UTILITIES

An important aspect of this study was to determine the applicability of using ClO_2 in combination with ozonation at the Randall-Bold WTP, but equally important was demonstrating the applicability for other utilities that use ozone or ClO_2. As with most testing and process changes, the data collected from this study are specific to the Randall-Bold WTP, but some general recommendations for other utilities include:

- For utilities that currently use ozone and must limit its use due to bromate formation, preoxidation with ClO_2 may offer an alternative oxidant that helps meet the treatment goals achieved by oxidation, such as enhanced filtration and DBP formation reduction, while minimizing bromate formation. This study showed that preoxidation with ClO_2 could be combined with ozonation while limiting bromate formation and still maintaining treated water quality goals. However, the utility must consider the chlorate that is formed when ClO_2 addition is followed by ozonation.
- Utilities should consider the operations impacts of using both ClO_2 and ozone. This study showed that daily operator attention increased because of ClO_2 generation monitoring, process sampling, compliance sampling, and the impact the interaction of chlorite and ozone had on ozone residual monitoring.

- Utilities using ozone should also consider conducting studies to optimize the ozone generation and ozone dose. Optimization efforts can reduce operating costs and overall energy use.
- For utilities that are considering ozonation and conducting pilot or bench-scale testing, preoxidation with ClO_2 should be incorporated into the testing protocol for evaluation. In addition to being a potential bromate mitigation strategy, preoxidation with ClO_2 can reduce the required ozonation dose, which can translate into smaller ozonation systems and lower capital costs.

FUTURE WORK

As with most investigations, this study answered a majority of the original questions, but it also generated new ideas and questions that could not be fully investigated. These new ideas and questions warrant future work. Based on the findings from this study, a recommendation for future work includes the following:

- Conducting controlled bench-scale studies to investigate the mechanism for bromate control when ClO_2 has been added. This investigation should include determine possible intermediates, effect on bromate formation pathways, and the impact of ClO_2 and chlorite on ozone demand.
- Conducting a filter evaluation with parallel conditions to verify preoxidation with ClO_2 or a combination of ClO_2 and ozone maintains similar filtered turbidity and particle counts when compared to ozone only.
- Further testing to verify the TOC removal and DBP formation potential are not affected by substituting ClO_2 preoxidation for preozonation.

REFERENCES

Aieta, E.M., and J.D. Berg. 1986. A Review of Chlorine Dioxide in Drinking Water Treatment. *Jour. AWWA*, 78: 62-72.

Amy, G.L. and M.S. Siddiqui. 1999. *Strategies to Control Bromate and Bromide. AWWA Research Foundation, Denver, Colo. American Water Works Association, 42 p.*

Anderson, E., M.A. Larson, B. Corona-Vasquez, and B.J. Marinas. 2001. A Comparative Study of the Inactivation Kinetics of Cryptosporidium Parvum and Bacillus Subtilis Spores with Chemical Disinfectants. *Proceedings AWWA Water Quality Technology Conference,* Nov. 11-15, 2001, Nashville, TN.

APHA, AWWA, and WEF (American Public Health Association, American Water Works Association, and Water Environment Federation).1998. *Standard Methods for the Examination of Water and Wastewater.* 20th ed.,Washington, D.C., APHA.

Aston, R.N. 1947. Chlorine Dioxide Use in Plants on the Niagara Border." *Jour. AWWA* 39:7:687-690

Bissonette, E. and S. Allgeier.1997. Summary of Disinfection Application Practices, Evaluated Through the Analysis of ICR Plant Schematics and Small Systems CPRE Data. Presented by the authors at the Disinfectant/Disinfection By-Products Rule meeting of FACA, Office of Ground Water and Drinking Water, Technical Support Center, Cincinnati, Ohio, February, 1997.

California State Department of Health Services. 1999. Chlorine Dioxide Byproducts in Drinking Water: A Fact Sheet for Renal Dialysis Water Suppliers. October 1999, 2 p. (Available from Department of Health Services, P.O. Box 947232 MS92, Sacramento, Calif. 94234-7320).

Dietrich, A.M., R.C. Hoehn, L.C. Dufresne, L.W. Buffin, D.M.C. Rashash, and B.C. Parker. 1995. "Oxidation of Odorous Algal Metabolites by Permanganate, Chlorine, and Chlorine Dioxide," Presented at the Fourth International Symposium on Off-Flavors in the Aquatic Environment, Adelaide, South Australia, October 2-7, 1994. *Water Science and Technology*, Vol. 13 (11): 223-228.

Dietrich, A.M., M.P. Orr, D.L. Gallagher, and R.C. Hoehn. 1992. Tastes and Odors Associated With Chlorine Dioxide, Jour. *AWWA*, Vol. 92: No. 6: 82-88, 1992.

EPRI 1999. "Issues for Ozone for Drinking Water Treatment", Electric Power Research Institute TechCommentary, TC-113030, Palo Alto, CA, 1999.

Federal Register, 1979. *National Interim Primary Drinking Water Regulations: Control of Trihalomethanes in Drinking Water Final Rule.* 40 CFR Parts 141 and 142, Vol.44 (No.231), 68624-6870, Nov. 29, 1974.

Federal Register. 1994a. *National Primary Drinking Water Regulations: Disinfectants and Disinfection Byproducts, Proposed Rule.* 40 CFR Parts 141 and 142, Vol. 59 (No. 145): 38668-38829, July 29, 1994

Federal Register. 1994b. *National Primary Drinking Water Regulations; Enhanced Surface Water Treatment Requirements; Proposed Rule,* 40 CFR Parts 141 and 142, Vol. 59, No. 145, 38832-38858, July 29, 1994.

Federal Register. 1996. *National Primary Drinking Water Regulations: Monitoring Requirements for Public Drinking Water Supplies; Final Rule.* 40 CFR Part 141, Vol. 61, No. 94, pp. 24354-24388, May 14, 1996.

Federal Register. 1998. *National Primary Drinking Water Regulations: Disinfectants and Disinfection Byproducts; Final Rule.* 40 CFR Parts 9, 141 and 142; Vol. 63 (No. 241): 69389-69476. December 16, 1998.

Finch, G.R., L.L. Gyürék, L.R.J. Liyanage, and M. Belosevic. 1997. *Effect of Various* Disinfection *Methods on the* Inactivation *of Cryptosporidium. AWWA Research Foundation, Denver, Colo. American Water Works Association, 66 p.*

Finch, G.R. and H. Li. 1999. Inactivation of *Cryptosporidium* at 1°C Using Ozone or Chlorine Dioxide, *Ozone Science & Engineering*, Vol. 21, No. 5, 477-496.

Gates, D. 1998. *The Chlorine Dioxide Handbook.* Eds. B. Cobban, P. Murray, and A. Livingston. Denver, Colo. AWWA, 186 p.

Gordon, G., R. G. Kieffer, and D.H. Rosenblatt. 1972, The Chemistry of Chlorine Dioxide. In: *Progress in Inorganic Chemistry.* Vol.15. New York, NY: Wiley-Interscience Publishers.

Gordon, G. 1992. Water Chemistry of the Oxy-Chlorine Species, *Chlorine Dioxide: Drinking Water Issues.* Second Int'l Symposium, Denver, Colo., American Water Works Assoc.

Gordon, G. and A. Rosenblatt. 1996. Gaseous, Chlorine-Free Chlorine Dioxide for Drinking Water, Proceedings 1995 Water Quality Technology Conference, New Orleans, La., November 12-16, 1995, Denver, Colo. AWWA. pp. 457-465.

Gordon, G. and B. Bubnis 1999. Ozone and Chlorine Dioxide: Similar Chemistry and Measurements Issues, *Ozone Science & Engineering*, Vol. 21, No. 5, 447-464.

Glaze, W.H., R. Schep, W. Chauncey, E.C. Turh, J.J. Zarnoch, E.M. Aieta, C.H. Tate, and M.J. McGuire. 1990. Evaluating Oxidants for the Removal of Model Taste and Odor Compounds, *Jour. AWWA,* 82: No. 5: 79-84.

Griese, M.H., K. Hauser, M. Berkemeier, and G. Gordon. 1991. Using Reducing Agents to Eliminate Chlorine Dioxide and Chlorite Ion Residuals in Drinking Water. *Journal AWWA,* 83:5:56-61.

Griese, M.H., J.J. Kaczur, and G. Gordon. 1992. Combining Methods for the Reduction of Oxychlorine Residuals in Drinking Water. *Journal AWWA*, 84:11:69-77.

Hoehn, R.C., A.M. Dietrich, W.S. Farmer, R.G. Lee, E.M. Aieta, D.W. Wood, and G. Gordon. 1990. Household Odors Associated With the Use of Chlorine Dioxide During Drinking Water Treatment, *Jour. AWWA,* Vol. 90: No. 4: 166-172.

Hoehn, R.C., B.W. Long, and D.J. Gates. 2000. Status of Chlorine Dioxide Disinfection Technologies, *Proceedings American Water Works Association Annual Conference, Denver, Colo. June 16-21, 2000.*

Hoehn, R.C., C.S. Ellenberger, D.L. Gallagher, E.V. Wiseman, R.W. Benninger, and A. Rosenblatt. [N.d.] Chlorine Dioxide and By-Product Persistence in a Drinking Water System. *Jour. AWWA*. Forthcoming.

Hoigné, J., H. Bader, W.R. Haag, and J. Staehelin. 1985. Rate Constants of Reactions of Ozone with Organic and Inorganic Compounds in Water - III, *Water Research,* Vol. 19: No. 9: 993-1004.

Hoigné, J. 1998. Chemistry of Aqueous Ozone and Transformation of Pollutants by Ozonation and Advanced Oxidation Processes. In: *The Handbook of Environmental Chemistry,* Vol. 5, Part C.: *Quality and Treatment of Drinking Water II.* Ed. J. Hrubec,, Berlin Heidelberg, Springer-Verlag.

Hurst, G.H., and W.R. Knocke. 1997. Evaluating Ferrous Iron for Chlorite Ion Removal. *Jour. AWWA*, 89:8:98-105.

Iatrou, A., and W.R. Knocke. 1992. Removing Chlorite by the Addition of Ferrous Iron. *Jour. AWWA*, 84:11:63-68 (November 1992)

Limoni, B., E.C. Goldstein, and C. Rav-Acha. 1984. Determination of Oxidants Formed Upon the Disinfection of Drinking Water with Chlorine Dioxide. *Jour. Environ. Sci. Health*, A19:8:943-957

Liyanage, L.R.J., G.R. Finch, and M. Belosevic. 1997a. Sequential Disinfection of *Cryptosporidium parvum* with Ozone and Chlorine Dioxide, *Ozone Science & Engineering*, Vol. 19, No. 5, 409-423.

Liyanage, L.R.J., L.L. Gyürék, M. Belosevic, and G.R. Finch. 1997b. "Effect of Chlorine Dioxide Preconditioning on Inactivation of *Cryptosporidium* by Free Chlorine and Monochloramine: Process Design Requirements. *Proceedings 1996 Water Quality Technology Conference*, Boston, Nov. 17-21, 1996, Denver, Colo., AWWA, 1997b.

Liyanage, I.R.J., G.R. Finch, and M. Belosevic. 1997c. Effect of Aqueous Chlorine and Oxychlorine Compounds on *Cryptosporidium parvum* Oocysts, *Environmental Science & Technology,* 31 (7): 1992-1994.

Long, B.W., R.A. Hulsey, and R.C. Hoehn. 1999. Complementary Uses of Chlorine Dioxide and Ozone for Drinking Water Treatment. *Ozone Science and Engineering,* Vol. 21, pp. 465-476.

Lykins, B.W., J.A. Goodrich, and J.C. Hoff. 1990. Concerns of Using ClO_2 Disinfection in the USA. *Jour. WaterSRT-Aqua,* 39:366-376.

Masschelein, W.J. 1979, *Chlorine Dioxide: Chemistry and Environmental Impact of Oxychlorine Compounds*. Ann Arbor, Mich., Ann Arbor Science Publishers

McCarthy, J.A. 1944. Chlorine Dioxide for the Treatment of Water Supplies. *Journal New England Water Works Association*, Vol. LVIII: 252-264.

McGuire, M.J., N.I. Lieu, and M.S. Pearthree. 1999. Using Chlorite Ion to Control Nitrification. *Jour. AWWA*, 91(20):52-61.

Minear, R.A. and G.L. Amy. 1996. Disinfection By-Products in Water Treatment. Lewis Publishers, New York, 1996, p.396.

Mitch, W.A. and D.L. Sedlak. 2002. Formation of N-Nitrosodimethylamine (NDMA) from Dimethylamine During Chlorination. *Environ. Sc. Technol.*, 36(4):588-595.

Noack, M.G. and R.L. Doerr. 1978. Reactions of Chlorine, Chlorine Dioxide, and Mixtures Thereof of Humic Acid: An Interim Report. In *Water Chlorination: Environmental Impact and Health Effects.* Vol. 2. Eds. Jolley, R.L., Gorchev, H., and Hamilton, D.H. Jr. Ann Arbor, Mich., Ann Arbor Science Publishers.

Pinkernell, U. and U. Von Guten. 2001. Bromate Minimization during Ozonation: Mechanistic Consideration. *Environ. Sci. Technol.*, 35:12:2424-2431.

Rav-Acha, C. 1984. The Reactions of Chlorine Dioxide with Aquatic Organic Materials and Their Health Effects. *Water Research.* 18:11:1329-134

Reckhow, D.A.1999. Control of Disinfection By-Product Formation Using Ozone. Chapter 9 in *Formation and Control of Disinfection By-Products in Drinking Water.* ed. P.C. Singer, Denver, Colo. AWWA, 424 pp.

Rice, R.G. 1996. Ozone Reference Guide, EPRI Document CR-106435. EPRI, Palo Alto, CA. April 1996.

Rice, R.G., P.K. Overbeck, and K. Larson. 1998. Ozone Treatment of Small Water Systems. *Proceedings 1998 IOA PAG Conference, Vancouver, British Columbia, Oct. 18-21, 1998.*

Rice, R.G., R.A. Hulsey, and J.J. Neemann. 2000. The ICR Bromate Story: What It May Mean for Ozone. *Proceedings 2000 IOA PAG Conference, Orlando, Florida, Oct. 1-4, 2000.*

Richardson, S.D. Identification of Drinking Water Disinfection By-Products. in: *Encyclopedia of Environmental Analysis and Remediation*. 1998. Vol. 3, pp. 1398-1421. Ed. Robert A. Meyers, John Wiley & Sons, Inc. New York, NY.

Summers, R.S., Hooper, S.M. and Shukairy, H.M. 1994 Development of Uniform Formation Conditions for the Assessment of Disinfection By-Product Formation. AWWA Annual Conference Proceedings, New York, New York June 19-23.

Synan, J.F., J.D. MacMahon, and G.P. Vincent. 1944. Chlorine Dioxide - Development in the Treatment of Potable Water. *Water Works and Sewer*, 91:12: 423-426.
United States Environmental Protection Agency (USEPA). 1997. *Community Water System Survey - Volumes I and II; Overview*. EPA 815-R-97-001a, -001b.

United States Environment Protection Agency (USEPA). 1999. Alternative Disinfectants and Oxidants Guidance Manual. EPA 815-R-99-014, Office of Water (4607), April 1999.

von Sonntag, C., G. Mark, R. Mertens, M.N. Schuchmann, and H.P. Schuchmann. 1993. UV Radiation and/or Oxidants in Water Pollution Control. *J. Water SRT – Aqua*. 42(4):201-211.

Wajon, J.E., D.HJ. Rosenblatt, and E.P. Burrows. 1982. Oxidation of Phenol and Hydroquinone by Chlorine Dioxide. *Environ. Sci. & Technol.*, 16:396-402

Werdehoff, K.S., and P.C. Singer. 1987. Chlorine Dioxide Effects on THMFP, TOXFP, and the Formation of Inorganic By-products. *Jour. AWWA*, 79:9:107-113.

Zegers, R.E., D.J. Rexing, E.C. Wert, T. Glew, J.J. Neemann, R.A. Hulsey, and B.W. Long. 2001. Bromate Formation and Mitigation While Achieving 2-log *Cryptosporidium* Inactivation on Lake Mead Water, Proceedings 2000 AWWA Annual Conference and Exhibition, Washington, D.C., June 12-16, 2001, Denver, Colo. AWWA.

Zika R.G., C.A. Moore, L.T. Gidel, and W.J. Cooper. 1985. Sunlight-Induced Photodecomposition of Chlorine Dioxide. pp. 1041-1053. In: *Water Chlorination: Chemistry, Environmental Impact, and Health Effects*. Vol. 5, eds. R.L. Jolley, R.J. Bull, W.P. Davis, S.Katz, M.H. Robert.

LIST OF ACRONYMS, ABBREVIATIONS, AND SYMBOLS

AL	Action Level
AWWA	American Water Works Association
AwwaRF	Awwa Research Foundation
Br^-	bromide ion
BrO_3^-	bromate ion
CCWD	Contra Costa Water Treatment
ClO_2	chlorine dioxide
CT	concentration multiplied by time
D/DBPR	Disinfectants/Disinfection Byproduct Rule
DBP	disinfection byproduct
DHS	Department of Health Services
GAC	granular activated carbon
gpm/ft^2	gallons per minute per square feet
HAA	haloacetic acid
ICR	Information Collection Rule
kW	kilowatt
kWh	kilowatt-hour
LGB	Lissamine Green B
MAC	Maximum Acceptable Concentration
MCL	Maximum Contaminant Level
MCLG	Maximum Contaminant Level Goal
MG	million gallons
mgd	million gallons per day
mg/L	milligrams per liter
MRDL	Maximum Residual Disinfectant Level
MRDLG	Maximum Residual Disinfectant Level Goal
MTBE	methyl tertiary-butyl ether
NDMA	N-Nitrosodimethylamine
NTU	Nephelometric Turbidity Unit
O_3	Ozone

ppb	parts per billion
RO	reverse osmosis
SDS	simulated distribution system
THM	trihalomethanes
TOC	total organic carbon
TOX	total organic halides
USEPA	United States Environmental Protection Agency
UV	ultraviolet light
UV254	ultraviolet absorbance at 254 nm
WTP	water treatment plant

IWA Publishing's authorised EU representative for General Product Safety Regulations is Diane D'Arras, 15 rue Duret, 75116 Paris, France, e-mail: safety@iwap.co.uk.

Printed and bound by CPI Group (UK) Ltd, Croydon, CR0 4YY
17/09/2026
02218804-0013